New Perspectives on Optimization Algorithms Research

Edited by

Germano Lambert-Torres ,
Gilberto Capistrano Cunha de Andrade
and Carlos Henrique Valério De Moraes

Published in London, United Kingdom

New Perspectives on Optimization Algorithms Research
http://dx.doi.org/10.5772/intechopen.1005697
Edited by Germano Lambert-Torres, Gilberto Capistrano Cunha de Andrade and Carlos Henrique Valério De Moraes

Contributors
Aljaž Jeromel, Blaž Repnik, Borut Žalik, Caliane B.B. Costa, Carlos Henrique Valério de Moraes, Chengkang Liu, Dongqiao Wang, Germano Lambert-Torres, Gilberto Capistrano Cunha de Andrade, Ivana Kolingerová, Leandro V. Pavão, Lingsong He, Longxiang Shan, Mauro A.S.S. Ravagnani, Niko Lukač, Qiaohua Wang, Ramadan Abdelaziz, Shaolong Shi, Shiwei Liu, Tiangui Hu, Xia Hua, Yanhua Sun, Yuandi Li

First published in London, United Kingdom, 2025 by IntechOpen
IntechOpen is the global imprint of INTECHOPEN LIMITED, registered in England and Wales, registration number: 11086078, 167-169 Great Portland Street, London, W1W 5PF, United Kingdom

For EU product safety concerns: IN TECH d.o.o., Prolaz Marije Krucifikse Kozulić 3, 51000 Rijeka, Croatia, info@intechopen.com or visit our website at intechopen.com.

British Library Cataloguing-in-Publication Data
A catalogue record for this book is available from the British Library

New Perspectives on Optimization Algorithms Research
Edited by Germano Lambert-Torres, Gilberto Capistrano Cunha de Andrade and Carlos Henrique Valério De Moraes

p. cm.
Print ISBN 978-1-83634-434-6
Online ISBN 978-1-83634-433-9
eBook (PDF) ISBN 978-1-83634-435-3

If disposing of this product, please recycle the paper responsibly.

Meet the editors

Germano Lambert-Torres received his Ph.D. in Electrical Engineering from the École Polytechnique de Montréal. He is a Professor at the Gnarus Institute. He has led over 100 R&D projects in intelligent systems, maintenance management, and power system operation for many power industries in Brazil and South America. He is the author of more than 25 patents and 80 registered software programs. He has taught seminars and short courses worldwide in numerous countries at conferences of IEEE, CIGRÉ, IFAC, universities, and research centers. He has supervised over 150 MSc and Ph.D. theses and published more than 800 journal and technical conference papers. He is also the author or editor of twelve books. Dr. Lambert-Torres is a Fellow of the IEEE and the AAIA.

Gilberto Capistrano Cunha de Andrade is the Tutoring Coordinator at the Education Center of Itajuba Federal University (UNIFEI), a Professor and Associate Researcher at the Gnarus Institute, and a Professor at the Federal Institute of Southern Minas. He holds a Ph.D. in Applied Mathematics and Statistics from the University of Beira Interior, a master's degree in Energy Engineering from the UNIFEI, specializations in Mathematics and Statistics from the Lavras Federal University and in Applied Geoprocessing from the Federal Institute of Northern Minas Gerais, and a full degree in Mathematics from the Faculty of Philosophy, Sciences and Languages of Itajuba. He served as Deputy Director at the Itajuba Professional Education Center. He participates in several R&D projects for Brazilian electricity companies.

Carlos Henrique Valerio de Moraes is a Full Professor at the Itajuba Federal University, where he also completed his doctorate, master's, and undergraduate degrees in Electrical Engineering. At this institution, he served as the Director of Information Technology. He specializes in automating electrical and industrial processes. He has participated in numerous research and development projects for industries and companies in the electricity sector, focusing primarily on artificial intelligence, intelligent systems, meta-heuristics, operation of electrical systems, smart controls, and processes to support decision-making and intelligent planning of actions. He has authored dozens of scientific articles, hundreds of software programs, and the book "Programming of Embedded Systems" in its second edition. He has supervised several undergraduate and graduate students.

Contents

Preface

In today's world, virtually every activity needs some optimization process. Often, these optimizations occur implicitly and almost naturally, from a slight modification in a simple method to some empirical changes in more sophisticated processes. However, when processes become complex, such as when there are many variables or external interferences, optimization processes must keep pace with this complexity.

Historically, complex processes were initially represented by arithmetic mathematical equations. With this came several optimization techniques, called mathematical (or classical), to manipulate these mathematical equations, aiming to maximize (or minimize) some objective function. It has generated a new field of knowledge: optimization algorithms.

Subsequently, another group of complex problems was added to the list of issues studied by the optimization algorithms, which were problems better represented by differential mathematical equations. As a result, another set of techniques has emerged to optimize this new class of problems.

Classical optimization techniques have been studied and disseminated for several decades, notably through the use of computational methods that have transformed their algorithms into increasingly powerful tools.

However, there was still a class of complex problems that could not be easily expressed by mathematical equations, whether in arithmetic or differential form. Classical optimization techniques could not be applied to this class of problems because there were no equations to express them. Algorithms, datasets, and implicit knowledge were used to describe these problems. Hence, heuristic techniques (and meta-heuristics) emerged to address this type of problem, providing optimization solutions for these problems.

With the advent of Artificial Intelligence techniques, they began to be applied to problems without an explicit mathematical formulation, with exciting results, especially when using techniques with biological or natural inspirations. Then, these intelligent techniques also began to compete with classical optimization techniques, optimizing problems with explicit mathematical formulations.

More recently, with the emergence of new computational structures on the market, they have also begun to contribute to optimization algorithms, making their algorithms and methods available to help solve complex problems more efficiently. Many of these techniques avoid the difficulties associated with classical and intelligent techniques, allowing them to arrive at better-quality solution points and further optimize the desired solution.

This book is a contribution to the area of Optimization Algorithms. It is divided into two main parts: Methodologies and Applications. The first part focuses on presenting modern optimization techniques, while the second part aims to encourage the application of these methods to various classes of problems.

The first part begins with a chapter entitled "Overview of the New Perspectives on Optimization Algorithms Research", which presents and classifies modern optimization techniques as they relate to the developments of Artificial Intelligence and quantum computers. The chapter begins with a brief classification of problems associated with optimization methods, and then it draws a parallel between classical and modern optimization techniques. The chapter classifies modern optimization algorithms into quantum mechanical and meta-heuristic techniques. Several subdivisions are presented and detailed throughout this chapter.

The second chapter of the first part is entitled "An Improved Particle Swarm Optimization Method for Nonlinear Optimization". It presents a modified meta-heuristic technique to solve a significant class of problems through nonlinear optimization concepts. This chapter begins with an interesting analysis of classical methods and then suggests modifications to the classical PSO. Studies of various nonlinear problems are presented, along with numerical results and comparative analyses.

The third chapter that complements the first part of this book is entitled "Bi-Level Non-deterministic Optimization Methods Applied in Process Synthesis". In this chapter, he presents an optimization analysis technique that applies when both integer and real variables are present. This technique is advantageous in terms of computational time for real-time solutions. It presents illustrative examples to demonstrate the advantages of the proposed method in a specific class of problems.

The second part of this book presents the application of optimization techniques to various problems, inspiring readers to apply these techniques to their problems.

The first chapter of this part is entitled "Introducing Global Optimization vs. Local Optimization in Water Resource Management". It compares global and local optimization techniques through studies of the Associated Environment. This problem is interesting because you want to maximize the use of available water resources while minimizing the negative impacts of that use. This problem is interesting because it represents a class of problems where multiple goals must be achieved. This chapter presents the results of studies, allowing the reader to verify which technique best applies to the solution of their problem.

The second chapter, entitled "Programming Optimisation of Move with Interleaving Transform", presents how to make efficient techniques that work with strings. It addresses various forms of implementation by comparing CPU times. To illustrate the proposal made in this chapter, various images are analyzed, demonstrating that the dynamically linked list offers the best performance, particularly when implemented using a standard vector.

The book's last chapter is called "Swarm Intelligence-Inspired Nanobiosensing for Tumor Microenvironment Targeting". This chapter presents an optimization technique for determining strategies that enhance the performance of nanorobots, which function as intelligent agents to solve problems jointly. The central idea of this chapter is to enable the reader to solve problems with intelligent multi-agents and suggest strategies for these agents.

The editors of this book hope that the presentation of these modern optimization techniques and their application examples will help readers apply them successfully, transferring them to other applications and areas.

Germano Lambert-Torres
Gnarus Institute,
Itajuba, Brazil

Gilberto Capistrano Cunha de Andrade
UAB/CEDUC/UNIFEI,
Itajuba, Brazil

Carlos Henrique Valério de Moraes
Itajuba Federal University,
Itajuba, Brazil

Section 1

Methodologies

Chapter 1

Overview of the New Perspectives on Optimization Algorithms Research

Germano Lambert-Torres, Carlos Henrique Valério de Moraes and Gilberto Capistrano Cunha de Andrade

Abstract

This chapter presents a brief overview of the most recent optimization algorithms. The chapter starts by showing the difference between classical methods and modern methods. And then, some special types of problems solved by optimization algorithms are presented, including some open problems. After that, the initial modern optimization algorithms are listed; the recent optimization algorithms are shown with their main features. This section starts with the algorithms inspired by quantum mechanics and their solutions to intelligent systems. Also, this section shows other modern approaches merging intelligent techniques, neuroscience, data science, and machine learning.

Keywords: quantum mechanics, warm intelligence-based algorithms, nature-inspired algorithms, gradient-based optimization algorithms, metaheuristic algorithms, hybrid optimization algorithms, reinforcement learning-based optimization, Bayesian optimization, neuroevolution, federated optimization, adversarial optimization, meta-learning for optimization

1. Introduction

Optimization is a fundamental process in several areas of knowledge, such as engineering, economics, computer science, and mathematics, aiming to find the best solution or the optimal point for a given problem. This process involves maximizing or minimizing an objective function, representing the metric to be improved, subject to constraints that delimit the space of viable solutions. Optimization can be applied in varied contexts, from improving industrial processes to efficiently allocating resources to complex projects [1].

There are different optimization techniques, which can be classified into classical methods and modern methods. Classical methods include linear programming, nonlinear programming, and differential calculus, which are widely used in continuous and differentiable problems. Using intelligent systems, modern techniques are often employed in complex, nonlinear, or multi-optimal problems. The choice of the appropriate method depends on the characteristics of the problem, such as the nature of the variables, the presence of constraints, and the complexity of the objective function [2, 3].

With the advancement of technology and the increase in computing capacity, optimization has become increasingly relevant, especially with the emergence of large volumes of data and the need to make quick and accurate decisions. Optimization tools are essential in applications such as machine learning, where models are tuned to minimize forecasting errors, or logistics, where routes and resources are planned to reduce costs and time. Optimization is crucial in the search for efficiency and innovation, contributing to developing more competent and sustainable solutions in an increasingly complex world [4].

2. Type of problems solved by optimization

Optimization has been widely used to solve various complex problems in different industries, providing efficient and cost-effective solutions. In logistics and transportation, for example, optimization is applied to determine the most effective routes, minimizing fuel costs, delivery time, and pollutant emissions. Optimization algorithms help plan the distribution of goods, vehicle allocation, and inventory management, ensuring that resources are used intelligently and sustainably. Companies like Amazon and FedEx utilize these techniques to improve the efficiency of their global operations.

In the field of energy, optimization plays a crucial role in the management of electricity grids and the integration of renewable sources. Problems such as energy allocation in smart grids, minimizing transmission losses, and balancing supply and demand are solved with optimization techniques. Additionally, optimization is essential in planning solar and wind power generation systems, where it is necessary to maximize energy production while considering factors such as weather conditions and installation costs. These applications contribute to the transition to a cleaner and more efficient energy matrix.

Another notable example of problems solved by optimization is in healthcare, where optimization techniques are used to improve the allocation of medical resources, the scheduling of surgeries, and the distribution of medications. In times of crisis, such as during the COVID-19 pandemic, optimization was key to planning vaccine distribution, hospital bed allocation, and medical team management. In addition, in medical research, optimization is applied in developing personalized treatments and analyzing genetic data, helping to identify patterns and solutions that improve patients' quality of life. These examples illustrate how optimization can be a powerful tool for addressing complex challenges and improving efficiency in various contexts.

2.1 Difficult problems to solve by optimization

Some problems are particularly challenging for optimization because of their complexity, scale, or dynamic nature. A typical example is the Traveling Salesman Problem (TSP). In this problem, a route is sought to find the shortest possible route to visit all cities and return to the starting point [5]. Although it seems simple, this problem becomes highly complex as the number of cities increases and the number of possible routes grows exponentially. Even with advanced algorithms, such as branch-and-bound or heuristics, finding the optimal solution for large TSP instances can be computationally infeasible, requiring approximate or metaheuristic approaches [6].

Another complex problem is optimizing chaotic or highly nonlinear systems, such as those in weather forecasts or financial markets. These systems are sensitive to slight variations in initial conditions, making it challenging to model and optimize them accurately. In addition, multiple local optimums and the interdependence between variables further complicate the search for global solutions. Techniques such as genetic algorithms, simulated annealing or other metaheuristic can help, but they still do not guarantee the best possible result, especially in highly uncertain scenarios.

Multiobjective optimization problems are also particularly challenging, as they involve maximizing or minimizing multiple objective functions that may be conflicting. For example, in the design of a car, one can seek to reduce the cost of production while maximizing safety and energy efficiency [7]. Finding an optimal balance between these goals is complex, as improving one aspect can degrade another. Methods such as goal programming or Pareto analysis are often used, but choosing the "optimal" solution frequently depends on subjective criteria, making decision-making difficult. These examples illustrate how optimization, while robust, faces limits when applied to intricate and multifaceted problems [8].

2.2 Some open problems by optimization

Despite its significant advances, optimization faces several open problems that challenge researchers and professionals. One of the critical challenges is the scalability of algorithms for large-scale problems [9]. With the exponential growth of data and the complexity of modern systems, many optimization algorithms become computationally infeasible for issues with millions of variables and constraints. This is particularly critical in areas, such as machine learning, where models need to be trained on large volumes of data, or transportation and logistics networks, where the number of possible routes and combinations is immense. Developing methods that maintain efficiency and accuracy globally is an issue that has not yet been entirely resolved.

Another open problem is optimization in dynamic and uncertain scenarios. Many real-world problems, such as managing supply chains, planning electricity grids, or allocating resources in real-time, involve constantly changing variables that are difficult to predict. The uncertainty associated with factors such as demand, costs, or environmental conditions makes traditional optimization insufficient, requiring adaptive and robust approaches. Despite advances in stochastic and robust optimization, much must be explored to ensure reliable and efficient solutions in highly volatile environments [10].

Finally, the integration of optimization with artificial intelligence (AI) and machine learning (ML) presents fascinating challenges that have not yet been fully solved. While ML techniques can enhance optimization algorithms, such as predicting parameters or reducing search spaces, combining these two areas is still a developing field. For example, how do you ensure that AI models generate optimized, interpretable, and generalizable solutions? How do we balance computational efficiency with precision in hybrid problems? These questions pave the way for interdisciplinary research that could revolutionize how we approach complex issues in the future.

3. Types of classical and modern optimization algorithms

Classical optimization algorithms form the basis for many modern techniques and are widely used in engineering, economics, logistics, and computer science problems. One of the best-known is the Simplex method, developed by George Dantzig in 1947 to solve linear programming problems [11]. Simplex is efficient in practice and allows you to find the optimal solution for problems with linear objective functions and constraints. Despite its effectiveness, the Simplex method has exponential complexity in the worst case, which has led to the development of alternatives, such as the interior point methods, which are more efficient for large-scale problems.

Another fundamental classical algorithm is the Gradient Descent method, which is widely used in continuous and nonlinear optimization problems. This iterative method seeks to minimize an objective function by following the direction of the most significant slope opposite the function's gradient. At the same time, simple and intuitive gradient descent can be slow in problems with complex cost surfaces, such as those with multiple local minima or narrow valleys. Variants such as the stochastic descending gradient (SGD) and the conjugate gradient have been developed to improve their efficiency and convergence, especially in machine learning applications.

Finally, Newton's method and its variants are essential for nonlinear optimization problems. Unlike the descending gradient, which uses only the first derivative, Newton's method uses the second derivative (Hessian) to achieve faster convergence, especially near the optimum. However, calculating Hessian can be computationally expensive, which has led to the development of quasi-Newton methods, such as the BFGS algorithm, that approximate Hessian without directly calculating second-order derivatives. These classic algorithms remain the backbone of many optimization applications as a starting point for developing more advanced techniques tailored to specific problems.

Modern optimization algorithms have gained prominence recently due to their ability to deal with complex, nonlinear, and large-scale problems, often inspired by natural phenomena or biological behaviors. One of the most important is the Particle Swarm Optimization (PSO) algorithm, which simulates the behavior of groups, such as flocks of birds or shoals of fish, to explore the solution-finding space. Each particle is a potential solution, and collaboration between them allows us to find global optimums on problems with multiple local minima. PSO is widely used in engineering, robotics, and machine learning due to its simplicity, easy implementation, and effectiveness.

Another recent and relevant algorithm is Optimization by Simulated Annealing (SA), inspired by the annealing process in metallurgy. This method allows for exploring the search space by temporarily accepting worse solutions, which helps people escape from significant locations. SA is beneficial in combinatorial optimization problems, such as the Traveling Salesman Problem, and in applications where the objective function is irregular or discontinuous. The flexibility and robustness of Simulated Annealing have made it a valuable tool in circuit design, logistics, and bioinformatics.

In addition, evolutionary algorithms, such as Genetic Algorithms (GA), continue to evolve and be applied to complex problems. Inspired by Darwin's theory of evolution, these algorithms use operations such as selection, crossbreeding, and mutation to become a population of solutions over generations. Recently, hybrid techniques that combine genetic algorithms with machine learning or deep learning have been developed to solve high-dimensional problems, such as optimizing hyperparameters in neural networks or designing autonomous systems. These modern methods represent the frontier of optimization, allowing the resolution of challenges considered intractable a few decades ago.

4. Recent optimization algorithms

Optimization constantly evolves, with new algorithms and techniques developed to solve complex problems more efficiently. Here are some of the latest and most promising optimization algorithms that have gained attention in recent years:

4.1 Optimization algorithms inspired by quantum mechanics

These algorithms work through tunneling and quantum superpositions, allowing for more efficient solution space exploration. These quantum effects help algorithms escape local minima more effectively than classical methods. They can also evaluate multiple solutions simultaneously.

Several quantum algorithms require quantum hardware, which is still in the early stages of development. Another challenge with these algorithms is that they may not scale well for significant problems. In addition, quantum systems are susceptible to noise and errors, which can affect performance.

Currently, the main lines of research are linked to developing more robust quantum algorithms for classical computers, integrating machine learning techniques, and exploring new quantum phenomena for optimization.

Optimization algorithms inspired by quantum mechanics represent a promising frontier for solving complex problems, especially as quantum computing technology advances.

4.1.1 Application of quantum mechanics with intelligent techniques

Several techniques unite the ideas of quantum mechanics with the intelligent techniques already applied to optimization problems.

Quantum Annealing: Inspired by quantum mechanics, this technique is used to find the global minimum of a function, taking advantage of quantum tunneling and superposition. This strategy was proposed by Kadowaki and Nishimori in 1998 [12]. The original proposal explored the idea of using quantum effects to optimize combinatorial problems, taking inspiration from the classical technique of Simulated Annealing but using quantum tunnels instead of thermal fluctuations. The company D-Wave Systems has been developing practical applications by building specialized hardware.

Quantum-Inspired Simulated Annealing: This algorithm, developed in Fujitsu's Japanese labs [13], combines classic simulated annealing with quantum tunneling effects. Instead of relying solely on thermal fluctuations to escape local minima, the algorithm incorporates quantum tunneling to explore the solutions space more effectively. This algorithm has proven helpful in optimization problems with rugged energy landscapes. Fujitsu introduced the method as part of its Digital Annealer technology, a system mimicking Quantum Annealing (QA) aspects without needing an actual quantum processor.

Quantum Genetic Algorithms (QGA): this algorithm, proposed by Narayanan and Moore [14], combines genetic algorithms with quantum computing principles such as superposition and entanglement. It uses quantum bits (qubits) to represent solutions, allowing the superposition of multiple states. Quantum gates are applied to evolve the population of solutions. It explores a more extensive solution space simultaneously due to quantum parallelism.

Quantum Particle Swarm Optimization (QPSO): This algorithm extends classical particle swarm optimization (PSO) by incorporating quantum behavior. He was initially proposed by Jun Sun, Bin Feng, and Wenbo Xu [15]. In this algorithm, the particles in the cluster are represented by wave functions, allowing them to exist in multiple states simultaneously. This allows for a more global exploration of the solutions space. In this way, the particles do not follow fixed trajectories, as in the traditional PSO, but evolve according to a probability distribution function, allowing a greater capacity to escape from local minima.

Quantum-Inspired Evolutionary Algorithm (QEA): This algorithm combines differential evolution with principles of quantum computing. This approach was proposed by Li and Li [16]. It uses mutation and crossover operators inspired by quantum mechanics to improve the exploration and exploitation of the solution space. It aims to improve the efficiency of the traditional evolutionary algorithm, especially in complex and multimodal problems, by incorporating quantum mechanisms such as superposition, interference, and quantum tunneling.

4.1.2 Other applications with quantum mechanics

There are also other applications with quantum mechanics, which are listed below.

Variational Quantum Algorithms (VQAs): this algorithm uses a hybrid quantum-classical approach to optimize a cost function. In it, a quantum circuit prepares a trial state, and then some adjustments are made by classical optimization methods of the circuit parameters to minimize the cost function. Several research groups have studied this technique, and it can be said that the pioneers were the group of Peruzzo [17] and Farhi [18]. These algorithms have been applied to combinatorial optimization, machine learning, and simulations in chemistry.

Quantum Walks for Optimization (QWO): these algorithms are analogous to the previous one, using classical random walks and operators with interference and superposition. The paper by Farhi and Gutmann [19] introduced the idea of continuous Quantum Walks and their potential for optimization. Later, Shenvi et al. [20] showed how discrete Quantum Walks could speed up search algorithms. These algorithms have been used in graph traversal, search problems, and optimization.

Quantum-Inspired Reinforcement Learning (QIRL): This algorithm combines reinforcement learning with quantum principles to improve decision-making in complex environments. Quantum states represent policies or actions, and quantum operations update policy. The pioneering paper in this area was written by Dong et al. [21], who proposed a value update algorithm inspired by state superposition and quantum parallelism. This chapter laid the groundwork for subsequent research in quantum-inspired reinforcement learning. Currently, these algorithms are used in robotics, games, and autonomous systems.

4.2 Other recent optimization algorithms

This section presents more research lines with algorithms developed by various research groups worldwide.

4.2.1 Swarm intelligence-based algorithms

These algorithms present a computing paradigm inspired by the collective behavior of decentralized and self-organizing natural systems. These systems demonstrate an impressive ability to solve complex problems without centralized control. These algorithms are particularly effective for global optimization, where the vast search space is multimodal. Some recent examples are:

- Gray Wolf Optimizer (GWO) [22]: Inspired by gray wolves' social hierarchy and hunting behavior, this algorithm is used for continuous optimization problems. The method proposes a mathematical model based on the hierarchy of the pack (α, β, δ, ω) and hunting behavior, including pursuit, siege, and attack on prey.

- Whale Optimization Algorithm (WOA) [23]: Stimulated by the hunting behavior of humpback whales, the technique is primarily known as "bubble-net feeding." The algorithm mimics these whales' "bubble feeding" strategy, which involves spiraling movement around the prey. Based on three phases of search: surrounding prey, spiral bubbles, and exploring the environment. It has been applied to various benchmark functions and engineering design problems.

- Salp Swarm Algorithm (SSA) [24]: Mimics the swarming behavior of salps in oceans. This algorithm is used to solve optimization problems. The SSA divides individuals into leaders and followers, where leaders explore the environment, and followers adjust their positions based on the leaders. Simple to implement and perform well on real engineering problems.

4.2.2 Nature-inspired algorithms

These optimization and search methods are a class of optimization and problem-solving methods based on natural processes and behaviors, such as biological evolution, swarm behavior, ecosystems, and physical phenomena. These algorithms mimic efficient strategies observed in nature to solve complex optimization and machine-learning problems. These methods are beginning to be applied in artificial intelligence, engineering, and data science because of their ability to find approximate solutions to issues that are difficult to model mathematically. Some recent examples are:

- Firefly Algorithm (FA) [25]: This algorithm is used for multimodal optimization based on the flashing behavior of fireflies and the attraction mechanism based on the intensity of the light they emit. The FA simulates how fireflies attract each other, whereas those with higher light intensity (better solutions) attract the others.

- Bat Algorithm (BA) [26]: Mimics by the echolocation behavior of bats, this algorithm is used for continuous optimization problems. It simulates how bats adjust the frequency and amplitude of their ultrasonic signals to locate prey and avoid obstacles. It uses frequency, pulse rate and loudness to change the movement of the solutions.

- Cuckoo Search (CS) [27]: Inspired by the brood parasitism of some cuckoo species, this algorithm is used for global optimization. The algorithm uses an exploration mechanism based on Lévy flights, which allows long jumps to avoid stagnation in optimal locations.

4.2.3 Gradient-based optimization algorithms

These algorithms are mathematical methods used to find minimum or maximum functions, exploring their derivatives to guide the optimization process. They are widely employed in machine learning and neural networks, where Gradient Descent (GD) and its variations, such as Stochastic Gradient Descent (SGD), Momentum, Adam, and RMSprop, play a key role in adjusting model weights. These algorithms work by iteratively updating the parameters in the opposite direction of the loss function gradient, seeking to minimize error. Its efficiency depends on factors such as the learning rate and the cost function choice, which are essential for effectively training complex models. Some recent examples are:

- AdamW [28]: Presents a modification to the Adam optimizer, separating the weight decay term from the gradient update. This approach aims to improve the regularization and overall performance of the model, especially in image classification tasks.

- Layer-wise Adaptive Moments optimizer for Batch training (LAMB) [29]: LAMB adapts the learning rate layer-wise for training large models like BERT. LAMB dynamically adjusts learning rates for each network layer, making it easy to train with large batches.

4.2.4 Metaheuristic algorithms

These algorithms are high-level optimization techniques designed to solve complex problems efficiently by exploring and exploiting the search space intelligently. They do not guarantee that the optimal solution will be found, but they offer high-quality approximations in a feasible time. These algorithms are inspired by natural phenomena, evolutionary processes, and physical dynamics and are widely used in combinatorial optimization, artificial intelligence, and engineering. Some recent examples are:

- Harmony Search (HS) [30]: Inspired by the musical improvisation process of jazz musicians. In the original study, the authors demonstrated the effectiveness of HS by applying it to the design of water distribution networks, highlighting its competitive performance with other optimization methods existing at the time.

- Teaching-Learning-Based Optimization (TLBO) [31]: Simulates the teaching-learning process in a classroom. This algorithm is used for continuous optimization problems. It is based on two main stages: (1) Teacher Phase: The teacher tries to raise the average level of the class by teaching students (improvement of the overall solution). (2) Student Phase: Students interact with each other, learning and improving their solutions. It does not require specific parameter adjustments other than the number of iterations and population.

4.2.5 Hybrid optimization algorithms

These algorithms combine different optimization techniques to exploit the search space better and improve efficiency in obtaining solutions. These algorithms integrate approaches such as heuristics, metaheuristics, and exact methods to balance exploration and exploration, overcoming individual limitations of each technique. Some recent examples are:

- Hybrid Genetic Algorithm-Particle Swarm Optimization (GA-PSO) [32]: Combines the strengths of GA and PSO to improve optimization performance. This integration results in an algorithm balancing exploration and solution space exploration. The hybrid method proved effective in reducing the convergence time and improving the accuracy of the trained model.

- Hybrid Firefly Algorithm-Genetic Algorithm (FA-GA) [33]: Merges the capabilities of FA with the exploitation capabilities of GA, resulting in more efficient solutions to product design problems. It combines the global exploration capability of the FA with the local exploration of the GA, providing a more comprehensive and refined search in the solutions space. The results indicate that H-FAGA outperforms traditional methods regarding computational accuracy and efficiency.

4.2.6 Reinforcement learning-based optimization

This algorithm uses reinforcement learning (RL) concepts to solve optimization problems, where an agent learns how to make decisions, interact with an environment, and receive rewards. Unlike traditional approaches, this method dynamically adapts to the problem, exploring and exploiting solutions intelligently. Some recent examples are:

- Deep Q-Networks (DQN) [34]: Uses reinforcement learning to optimize decision-making processes in complex environments. Introduced the replay experience to store and reuse past samples, avoiding unwanted temporal correlations. It uses a fixed target network to stabilize training.

- Proximal Policy Optimization (PPO) [35]: A policy gradient method for optimizing reinforcement learning policies. Introduction of an objective function that uses a clipping to avoid overly drastic policy updates, improving training stability. PPO combines strong performance with ease of adjustment, making it a standard choice for many reinforcement learning problems.

4.2.7 Bayesian optimization

These algorithms use probabilistic optimization that manufactures probabilistic models to estimate the objective function and effectively determine the points to be evaluated, minimizing the number of evaluations required. It benefits problems with expensive objective functions, such as in machine learning models' hyperparameter tuning or engineering experiments. Rather than randomly or systematically testing the input values, Bayesian optimization utilizes an iterative approach, where the function is probabilistically modeled, and the subsequent evaluation is chosen based on uncertainty about the function, balancing exploration and exploration of the search space. Some recent examples are:

- Gaussian Process-Based Bayesian Optimization [36]: Gaussian processes are used to model the objective function and find the optimal solution with fewer evaluations. Introduction of acquisition functions such as Expected Improvement (EI) to balance exploration and exploitation efficiently. It demonstrates that Bayesian Optimization can find optimal settings for neural networks and SVMs more efficiently than grid search.

- Tree-Structured Parzen Estimators (TPE) [37]: A Bayesian optimization technique that uses tree-structured models to guide the search process. Unlike the Gaussian Processes used in traditional Bayesian Optimization, TPE models the conditional distribution of input values separately for good and bad solutions. Instead of assuming an objective function as a Gaussian process, TPE uses nonparametric models (Parzen Estimators) to estimate distributions of hyperparameters.

4.2.8 Neuroevolution

These algorithms combine the principles of neuroscience and biological evolution to train artificial neural networks. Inspired by natural selection, the method uses evolutionary algorithms to optimize neural network structure and parameters, such as weights and connections, over generations. Rather than relying solely on traditional learning methods such as backpropagation, neuroevolution generates a population of neural networks, evaluates their performance on a specific task, and selects the best ones for reproduction and mutation. This approach is instrumental in complex problems, such as robot control, gaming, and system optimization, where exploring diverse solutions can lead to groundbreaking discoveries. Neuroevolution stands out for its flexibility, allowing the adaptation of neural architectures dynamically and autonomously. Some recent examples are:

- NeuroEvolution of Augmenting Topologies (NEAT) [38]: An evolutionary algorithm that evolves neural network weights and topology. Unlike traditional machine learning methods, NEAT evolves not only the weights of connections but also the structure of the neural network. Introducing a speciation mechanism keeps different architectures evolving separately, preventing premature convergence. New nodes and connections are added gradually, allowing evolution to move away from simple networks and increase complexity only when necessary.

- CoDeepNEAT [39]: An extension of NEAT that evolves deep neural networks with complex architectures. The architecture evolves modularly, combining the evolution of neural layer blocks and the overall network structure. Based on the hierarchical evolutionary algorithm, smaller components can be developed separately and combined to form more complex architectures. It demonstrated strong performance on problems such as image classification and reinforcement learning, an essential step toward automated NAS (Neural Architecture Search).

4.2.9 Federated optimization

Federated Optimization is a distributed machine learning paradigm that allows you to train models collaboratively without centralizing data in a single location. In this approach, multiple devices or clients, such as smartphones or local servers, participate in the training process using their data, which remains stored locally.

Instead of sharing raw data, only model updates, such as gradients or parameters, are sent to a central server for aggregation. This preserves data privacy and reduces the risk of leaks, and it is especially relevant in industries such as healthcare and finance. However, federated optimization faces challenges such as data heterogeneity, communication latency, and device synchronization. Secure aggregation (e.g., Federated Averaging) and data compression methods are often employed to improve this approach's efficiency, scalability, and implementation. Some recent examples are:

- Federated Averaging (FedAvg) [40]: A distributed optimization algorithm used in federated learning to train models across decentralized devices. Instead of calculating gradients locally and sending them to the server, as in distributed SGD, FedAvg trains models locally for multiple iterations and then aggregates the updated models by weighted average. Dramatically reduces communication volume between client devices and the central server, making federated learning feasible in bandwidth-constrained scenarios.

- Federated Stochastic Variance Reduced Gradient (FSVRG) [41]: A variant of SVRG adapted for federated learning settings. The method improves training stability in scenarios where data distributed between clients does not follow the same distribution (non-IID). FedSVRG can outperform FedAvg in terms of efficiency and convergence, especially when there are significant variations in the data from client devices. FedSVRG and its variants have improved federated learning in deep neural networks, making training more efficient in mobile, IoT, and decentralized networks.

4.2.10 Adversarial optimization

In these algorithms, the core concept is based on machine learning techniques that involve competition between two neural models, as in the case of Generative Adversarial Networks (GANs). In this scenario, two agents are trained simultaneously: one is the generator, which creates synthetic data, and another is the discriminator, which tries to distinguish between accurate and generated data. The process is formulated as a zero-sum game. In this approach, the generator tries to minimize the discriminator's ability to correctly identify samples while the discriminator attempts to maximize its accuracy. This adversarial optimization leads to an equilibrium known as the Nash point, where the generator produces data indistinguishable from the actual data. Despite its power, adversarial optimization faces challenges such as mode collapse, training instability, and difficulty in convergence. Applications include imaging, style translation, data augmentation, and even security in deep learning models, where adversarial attacks are simulated to improve the robustness of systems. Some recent examples are:

- Generative Adversarial Networks (GANs) [42]: Use a minimax approach between a generator and a discriminator to optimize data generation. It showed that GANs can generate realistic images without an explicit loss function to measure the quality of the generated samples. GANs have become one of the most influential technologies in deep learning, and they are applied in image generation, videos, speech synthesis, art, and deepfakes. The concept of adversarial learning has inspired techniques in various areas, including safety and reinforcement learning.

- Adversarial Training [43]: A technique used to optimize models to be robust against adversarial attacks. It demonstrated that neural networks are vulnerable to small, imperceptible perturbations in the input data, which can lead to erroneous predictions. He proposed to train models by including adversarial examples in the training set, making them more robust. It showed that adversarial examples trained on one model can fool different models, indicating a fundamental vulnerability of deep neural networks.

4.2.11 Multiobjective optimization algorithms

These algorithms are techniques designed to solve problems that involve the simultaneous optimization of multiple, often conflicting, targets and objectives. Unlike traditional optimization, which seeks a single optimal solution, these algorithms aim to find a set of compromise solutions, known as the Pareto Front, where no goal can be improved without worsening another. These algorithms are widely applied in engineering, finance, logistics, and design, where decisions need to balance trade-offs between cost, performance, efficiency, and other criteria. The choice of method depends on the problem's complexity, the objectives' nature, and the need for diversity and precision in solutions. Some recent examples are:

- Non-Dominated Sorting Genetic Algorithm II (NSGA-II) [44]: is a multiobjective genetic algorithm designed to optimize multiple objectives simultaneously, balancing diversity and convergence. The algorithm introduces the concept of classification by non-domination to organize the population of solutions, representing the trade-off between the objectives. It uses elitism, maintaining the best solutions between generations, and crowding distance to preserve the diversity of solutions on each Pareto front.

- Multiobjective Particle Swarm Optimization (MOPSO) [45]: uses an archive of unmastered solutions to store the best solutions found, avoiding duplicates and maintaining a good diversity in the solutions. MOPSO improves PSO to balance convergence to Pareto fronts with solution diversity using particle mapping techniques to represent multiple objectives.

4.2.12 Hyperparameter optimization

In these algorithms, a process of tuning the model hyperparameters to improve its performance on specific tasks. Hyperparameters are parameters not learned directly from the data, such as the rate of learning, the number of layers in a neural network, or the depth of a decision tree. Hyperparameter optimization is essential, as the right choice can significantly improve model accuracy and generalizability. These algorithms seek to find the optimal combination of hyperparameters efficiently, considering the computational cost and training time. Some recent examples are:

- Hyperband [46]: This hyperparameter optimization algorithm based on bandit strategies aims to find the best parameter setting efficiently, even in large-scale problems. The main idea behind Hyperband is to allocate resources adaptively to different hyperparameter configurations. The algorithm starts with several configurations and dynamically allocates resources (such as training time or number of iterations), quickly discarding the least promising configurations.

- Bayesian Optimization and Hyperband (BOHB) [47]: Combines the efficiency of Hyperband (a resource allocation method based on successive halvings) with the probabilistic modeling capability of Bayesian Optimization. This combination allows for a more efficient and robust search for hyperparameters. BOHB is designed to be compute-efficient and scalable to large hyperparameter spaces. It utilizes Hyperband to quickly explore promising configurations and Bayesian Optimization to refine the search based on a probabilistic model.

4.2.13 Meta-learning for optimization

These algorithms, also known as "learning to learn," seek to improve optimization algorithms' performance over time by adapting to different problems or domains. Rather than treating each optimization problem as an isolated task, meta-learning uses past experiences to extract patterns and heuristics that can be applied to new problems. This is done by training a metamodel, such as a neural network or a reinforcement learning algorithm, to predict efficient hyperparameter configurations, initializations, or even search strategies. These algorithms are beneficial in scenarios where optimization problems are complex, dynamic, or have similar characteristics, such as in hyperparameter tuning of machine learning models, robot control, or resource planning. By reducing the time and resources required to find optimal solutions, meta-learning promises greater efficiency and scalability in practical applications. Some recent examples are:

- Model-Agnostic Meta-Learning (MAML) [48]: This meta-learning algorithm allows neural network models to quickly adapt to new tasks with few examples. The approach is model-agnostic, which can be applied to any neural network model. The MAML approach involves optimizing a loss function composed of a training and adaptation cycle: the model is trained to minimize error across multiple tasks and then is quickly adjusted for each task with a limited number of gradient updates.

- Reptile [49]: is a meta-learning algorithm designed to be simpler and more scalable than Model-Agnostic Meta-Learning (MAML). It combines quickly adapting to new tasks with a simplified second-order gradient process. Instead of using second-order gradient optimization, as in MAML, Reptile uses a first-order gradient-based algorithm, which makes the method more efficient and scalable, especially for problems of greater complexity.

5. Conclusions

The list of presented algorithms is not incomplete. Many groups are working on new algorithms, and some publications have not yet been printed. These algorithms are applied across various domains, including machine learning, operations research, engineering, and finance, to solve increasingly complex optimization problems. The choice of algorithm often depends on the specific characteristics of the problem, such as the nature of the objective function, constraints, and the available computational resources.

With the advent of quantum computing, hybrid classical-quantum algorithms are gaining prominence. These algorithms combine the ability of quantum computers to explore complex search spaces with the efficiency of classical methods to adjust parameters. As quantum hardware matures, these algorithms are expected to solve problems intractable to classical methods such as financial portfolio optimization and molecule simulation.

Integrating machine learning (ML) techniques and neural networks with optimization algorithms is revolutionizing the field. The combination of ML and optimization will enable the creation of adaptive algorithms that learn from historical data and dynamically adjust to new scenarios.

Large-scale optimization problems in big data, communication networks, and global supply chains require scalable and efficient algorithms. Distributed and parallelized algorithms will be essential to address the exponential data growth and modern systems' complexity.

Real-world problems often involve multiple conflicting goals and complex constraints. Advances in techniques for balancing objectives and constraints will allow for more balanced and practical solutions to complex problems.

Nature-inspired algorithms will continue to evolve, incorporating more sophisticated concepts from biology and ecology to enhance exploration and search within the search space.

Real-time optimization algorithms must be highly efficient and able to handle uncertainties and dynamic environmental changes. The demand for algorithms that operate in real-time is growing, especially in applications such as (a) traffic control and logistics in the dynamic optimization of routes and resources and (b) autonomous systems in decision-making for autonomous vehicles and drones.

Some future challenges are in the scalability of problems, dealing with increasingly more significant and complex issues, robustness, ensuring that algorithms work well in scenarios with uncertainties and noise, and interpretability, developing algorithms that provide interpretable and explainable solutions.

Another important aspect is the integration with new hardware, which takes advantage of advances in hardware, such as GPUs, TPUs, and quantum computers.

The future of optimization algorithms is promising, with advances in quantum computing, artificial intelligence, and nature-inspired techniques. As real-world problems become more complex, optimization algorithms must be more adaptive, scalable, and efficient.

Acknowledgements

The authors thank CNPq, CAPES, FAPEMIG, and the Brazilian Electricity Regulatory Agency Research and Development (ANEEL R&D) for supporting this project.

Conflict of interest

The authors declare no conflict of interest.

DOI: http://dx.doi.org/10.5772/intechopen.1010958

Author details

Germano Lambert-Torres[1*], Carlos Henrique Valério de Moraes[2]
and Gilberto Capistrano Cunha de Andrade[1]

1 Gnarus Institute, Itajuba, Brazil

2 Itajuba Federal University, Itajuba, Brazil

*Address all correspondence to: germanoltorres@gmail.com

References

[1] Gharbi M, Gharesifard B, Ebenbauer C. Anytime proximity moving horizon estimation: Stability and regret. IEEE Transactions on Automatic Control. 2023;**68**(6):3393-3408. DOI: 10.1109/TAC.2022.3190044

[2] Zhong S, Zhao X, Liu D, Su H, Xie Z, Fan B. HDSR: Image super-resolution method for harmonic diffraction optical imaging system based on plug and play technology. IEEE Transactions on Geoscience and Remote Sensing. 2025;**63**:1-19. DOI: 10.1109/TGRS.2025.3542150

[3] Watabe Y, Shibuya T. Nonlinear value function approximation method with easy hyperparameter tuning and convergence guarantee. IEEE Access. 2025;**13**:30117-30126. DOI: 10.1109/ACCESS.2025.3540900

[4] Hagebring F, Farooqui A, Fabian M, Lennartson B. On optimization of automation systems: Integrating modular learning and optimization. IEEE Transactions on Automation Science and Engineering. 2022;**19**(3):1662-1674. DOI: 10.1109/TASE.2022.3144230

[5] Ramirez A, Vellis V. Time complexity of the analyst's traveling salesman algorithm. Journal of Logic and Analysis. 2024;**16**:2. DOI: 10.4115/jla.2024.16.2

[6] Angelelli E, Bazgan C, Grazia Speranza M, Tuza Z. Complexity and approximation for traveling salesman problems with profits. Theoretical Computer Science. 2014;**531**:54-65. DOI: 10.1016/j.tcs.2014.02.046

[7] Sharifi MR, Akbarifard S, Qaderi K, et al. A new optimization algorithm to solve multiobjective problems. Scientific Reports. 2021;**11**:20326. DOI: 10.1038/s41598-021-99617-x

[8] Nand R, Chaudhary K, Sharma B. Discrete preference-based stepping ahead firefly algorithm for solving multidimensional knapsack problems. IEEE Access. 2024;**12**:139154-139164. DOI: 10.1109/ACCESS.2024.3466149

[9] Pardalos PM, Migdalas A, editors. Open Problems in Optimization and Data Analysis. 1st ed. Cham, Switzerland: Springer Cham; 2018. 330 p. DOI: 10.1007/978-3-319-99142-9

[10] Toloo M, Talatahari S, Rahimi I, editors. Multiobjective Combinatorial Optimization Problems and Solution Methods. 1st ed. Cambridge, USA: Academic Press; 2022. 290 p. DOI: 10.1016/C2020-0-00431-7

[11] Dantzig GB. Reminiscences about the origins of linear programming. Operations Research Letters. 1982;**1**(2):43-48. DOI: 10.1016/0167-6377(82)90043-8

[12] Kadowaki T, Nishimori H. Quantum annealing in the transverse Ising model. Physical Review E. 1998;**58**:5355-5363. DOI: 10.1103/PhysRevE.58.5355

[13] Sao M, Watanabe H, Musha Y, Utsunomiya A. Application of digital annealer for faster combinatorial optimization. Fujitsu Scientific & Technical Journal. 2019;**55**(2):45-51

[14] Narayanan A, Moore M. Quantum-inspired genetic algorithms. In: Proceedings of IEEE International Conference on Evolutionary Computation (CEC 1996); 20-22 May 1996; Nagoya, Japan. New York: IEEE; 1996. pp. 61-66. DOI: 10.1109/ICEC.1996.542334

[15] Sun J, Feng B, Xu W. Particle swarm optimization with quantum behavior.

In: Proceedings of the 2004 Congress on Evolutionary Computation (CEC 2004); 19-23 June 2004; Portland, USA. Vol. 1. New York: IEEE; 2004. pp. 325-331. DOI: 10.1109/CEC.2004.1330875

[16] Li P, Li S. Quantum-inspired evolutionary algorithm for continuous space optimization based on Bloch coordinates of qubits. Neurocomputing. 2008;**72**:581-591. DOI: 10.1016/j.neucom.2007.11.017

[17] Peruzzo A, McClean J, Shadbolt P, et al. A variational eigenvalue solver on a photonic quantum processor. Nature Communications. 2014;**5**:4213. DOI: 10.1038/ncomms5213

[18] Farhi E, Goldstone J, Gutmann S. A Quantum Approximate Optimization Algorithm. Cambridge, USA: Massachusetts Institute of Technology (MIT); 2014. DOI: 10.48550/arXiv.1411.4028

[19] Farhi E, Gutmann S. Quantum computation and decision trees. Physical Review A. 1998;**58**(2):915-928. DOI: 10.1103/PhysRevA.58.915

[20] Shenvi N, Kempe J, Whaley KB. Quantum random-walk search algorithm. Physical Review A. 2003;**67**(5):052307. DOI: 10.1103/PhysRevA.67.052307

[21] Dong D, Chen C, Li H, Tarn T-J. Quantum reinforcement learning. IEEE Transactions on Systems, Man, and Cybernetics, Part B. 2008;**38**(5):1207-1220. DOI: 10.1109/TSMCB.2008.925743

[22] Mirjalili S, Mirjalili SM, Lewis A. Grey wolf optimizer. Advances in Engineering Software. 2014;**69**:46-61. DOI: 10.1016/j.advengsoft.2013.12.007

[23] Mirjalili S, Lewis A. The whale optimization algorithm. Advances in Engineering Software. 2016;**95**:51-67. DOI: 10.1016/j.advengsoft.2016.01.008

[24] Mirjalili S, Gandomi AH, Zahra Mirjalili SZ, Saremi S, Faris H, Mirjalili SM. Salp swarm algorithm: A bio-inspired optimizer for engineering design problems. Advances in Engineering Software. 2017;**114**:163-191. DOI: 10.1016/j.advengsoft.2017.07.002

[25] Yang XS. Firefly algorithms for multimodal optimization. In: Watanabe O, Zeugmann T, editors. Stochastic Algorithms: Foundations and Applications, Lecture Notes in Computer Science. Vol. 5792. Berlin, Heidelberg: Springer; 2009. pp. 169-178. DOI: 10.1007/978-3-642-04944-6_14

[26] Yang XS. A new metaheuristic bat-inspired algorithm. In: González JR, Pelta DA, Cruz C, Terrazas G, Krasnogor N, editors. Nature Inspired Cooperative Strategies for Optimization. Studies in Computational Intelligence. Vol. 284. Berlin, Heidelberg: Springer; 2010. pp. 65-74. DOI: 10.1007/978-3-642-12538-6_6

[27] Yang X-S, Deb S. Cuckoo search via lévy flights. In: Proceedings of the 2009 World Congress on Nature & Biologically Inspired Computing (NaBIC); 9-11 December 2009; Coimbatore, India. New York: IEEE; 2009. pp. 210-214. DOI: 10.1109/NABIC.2009.5393690

[28] Loshchilov I, Hutter F. Decoupled weight decay regularization. In: Proceedings of the International Conference on Learning Representations (ICLR 2019); 6-9 May 2019; New Orleans, United States. Appleton, USA: ICLR. DOI: 10.48550/arXiv.1711.05101

[29] You Y, Li J, Reddi S, Hseu J, Kumar S, Bhojanapalli S, et al. Large batch optimization for deep learning: Training BERT in 76 minutes. In:

Proceedings of the Eighth International Conference on Learning Representations (ICLR 2020); 26 April-1st May 2020; Virtual Conference. Appleton, USA: ICLR. DOI: 10.48550/arXiv.1904.00962

[30] Geem ZW, Kim JH, Loganathan GV. A new heuristic optimization algorithm: Harmony search. Simulation. 2001;**76**(2):60-68. DOI: 10.1177/003754970107600201

[31] Rao RV, Savsani VJ, Vakharia DP. Teaching–learning-based optimization: A novel method for constrained mechanical design optimization problems. Computer-Aided Design. 2011;**43**(3):303-315. DOI: 10.1016/j.cad.2010.12.015

[32] Farag WA, Quintana VH, Lambert-Torres G. A genetic-based neuro-fuzzy approach for modeling and control of dynamical systems. IEEE Transactions on Neural Networks. 1998;**9**(5):756-767. DOI: 10.1109/72.712150

[33] Hafsi N, Hachimi H, Benterki D. A new hybrid genetic with firefly algorithm for solving cyber-criminal's attacks. In: Proceedings of the 2022 8th International Conference on Optimization and Applications (ICOA 2022); 6-7 October 2022; Genoa, Italy. New York: IEEE; 2022. DOI: 10.1109/ICOA55659.2022.9934759

[34] Mnih V, Kavukcuoglu K, Silver D, Graves A, Antonoglou I, Wierstra D, et al. Playing Atari with Deep Reinforcement Learning. DeepMind Technologies. Seattle: OpenTrain AI; 2013. DOI: 10.48550/arXiv.1312.5602

[35] Schulman J, Wolski F, Dhariwal P, Radford A, Klimov O. Proximal Policy Optimization Algorithms.DeepMind Technologies. Seattle: OpenTrain AI; 2017. DOI: 10.48550/arXiv.1707.06347

[36] Rasmussen CE, Williams CKI. Gaussian Processes for Machine Learning. 1st ed. Massachusetts, USA: The MIT Press; 2006. 248 p. DOI: 10.7551/mitpress/3206.001.0001

[37] Bergstra J, Bardenet R, Bengio Y, Kégl B. Algorithms for Hyper-Parameter Optimization. In: Proceedings of the 25th International Conference on Neural Information Processing Systems (NeurIPS 2011); 12-15 December 2011; Granada, Spain. Red Hook, NY: Curran Associates Inc.; 2011. pp. 2546-2554. DOI: 10.5555/2986459.2986743

[38] Stanley KO, Miikkulainen R. Evolving neural networks through augmenting topologies. Evolutionary Computation. 2002;**10**(2):99-127. DOI: 10.1162/106365602320169811

[39] Miikkulainen R, Liang J, Meyerson E, Rawal A, Fink D, Francon O, et al. Evolving Deep Neural Networks. The University of Texas at Austin. San Francisco: Sentient Technologies, Inc.; 2019. DOI: 10.48550/arXiv.1703.00548

[40] McMahan B, Moore E, Ramage D, Hampson S, y Arcas BA. Communication-efficient learning of deep networks from decentralized data. In: Proceedings of the 20th International Conference on Artificial Intelligence and Statistics (AISTATS 2017); 20-22 April 2017; Fort Lauderdale, USA. Seattle: Google, Inc.; 2017. pp. 1273-1282. DOI: 10.48550/arXiv.1602.05629

[41] Reddi SJ, Charles Z, Zaheer M, Garrett Z, Rush K, Konečný J, et al. Adaptive Federated Optimization, 2020. In: Proceedings of the Ninth International Conference on Learning Representations (ICLR 2021); 4-7 May 2021; Vienna, Austria. Appleton, USA: ICLR; DOI: 10.48550/arXiv.2003.00295

[42] Goodfellow I, Pouget-Abadie J, Mirza M, Xu B, Warde-Farley D, Ozair S, et al. Generative adversarial networks.

In: Proceedings of the International Conference on Neural Information Processing Systems (NIPS 2014); 8-13 December 2014; Montreal, Canada. San Diego: NeurIPS; 2014. pp. 2672-2680. DOI: 10.48550/arXiv.1406.2661

[43] Szegedy C, Zaremba W, Sutskever I, Bruna J, Erhan D, Goodfellow I, et al. Intriguing properties of neural networks. In: Proceedings of the International Conference on Learning Representations (ICLR 2014); 14-16 April 2014; Banff, Canada. Banff, Canada. Appleton, USA: ICLR; 2014. DOI: 10.48550/arXiv.1312.6199

[44] Deb K, Pratap A, Agarwal S, Meyarivan T. A fast and elitist multiobjective genetic algorithm: NSGA-II. IEEE Transactions on Evolutionary Computation. 2002;**6**(2):182-197. DOI: 10.1109/4235.996017

[45] Coello CAC, Lechuga MS. MOPSO: A proposal for multiple objective particle swarm optimization. Evolutionary Computation. 2002;**10**(3):353-377. DOI: 10.1109/CEC.2002.1004388

[46] Li L, Jamieson KG, DeSalvo G, Narayan S. Hyperband: A novel bandit-based approach to hyperparameter optimization. Journal of Machine Learning Research. 2018;**18**(1):6765-6816. DOI: 10.48550/arXiv.1603.06560

[47] Falkner S, Klein A, Hutter F. BOHB: Robust and Efficient Hyperparameter Optimization at Scale. In: Proceedings of the 35th International Conference on Machine Learning (ICML 2018); 10-15 July 2018; Stockholm Sweden. San Diego: NeurIPS; 2018. pp. 1437-1446. DOI: 10.48550/arXiv.1807.01774

[48] Finn C, Abbeel P, Levine S. Model-agnostic meta-learning for fast adaptation of deep networks. In: Proceedings of the 34th International Conference on Machine Learning (ICML 2017); 6-11 August 2017; Sydney, Australia. San Diego: NeurIPS; 2017. pp. 1126-1135. DOI: 10.48550/arXiv.1703.03400

[49] Nichol A, Schulman J. Reptile: A Scalable Metalearning Algorithm. Seattle: OpenTrain AI; 2018. DOI: 10.48550/arXiv.1803.02999v1

Chapter 2

An Improved Particle Swarm Optimization Method for Nonlinear Optimization

Shiwei Liu, Yuandi Li, Longxiang Shan, Chengkang Liu, Dongqiao Wang, Qiaohua Wang, Xia Hua, Yanhua Sun and Lingsong He

Abstract

Nonlinear optimization is becoming challengeable in information sciences and various industrial applications, but nonlinear problems solved by the classical particle swarm-based methods are usually characterized with low efficiency, accuracy, and convergence speed in specific issues. To solve these problems and enhance the nonlinear optimization performance, an improved metaheuristic particle swarm optimization (PSO) model is proposed here. First, the optimization principles and model of the new method are introduced, and algorithms of the improved PSO are presented. Then, influence of the model parameters, input dimensions, and different nonlinear problems on the PSO optimization characterizations are studied by Pareto set solving and optimization performance comparison. The analysis regarding diverse nonlinear problems and optimization methods demonstrates the feasibility of the proposed method. Finally, the performance evaluation is exhibited by the case study of nonlinear parameter optimization, CEC benchmark problems, and rank sum test, which all verify its effectiveness and reliability, as well as the significance and great application promise. Additionally, the future work is discussed.

Keywords: particle swarm optimization (PSO), nonlinear optimization, performance evaluation, Pareto set, benchmark problem

1. Introduction

Nonlinear optimization problems including the constrained and unconstrained situations have drawn more and more attention, where the law of physics is usually formulated as the mathematical form and theories of maximum or minimum of the objective functions. Thus, optimal solutions can be obtained, and the performance of objective functions or the model to be optimized can be enhanced. The nonlinear optimization techniques may include the steepest decent, Newton method, conjugate gradient, trust region, non-gradient methods for unconstrained problems, and the basic dual, feasible point, penalty, augmented Lagrangian, and barrier methods for constrained problems [1], which have been widely applied in engineering, economics, portfolio management, medical diagnosis, computer science, and mathematics when the theories of statistics, approximation and operational research are considered [1, 2]. Typical applications and studies are reported in keyframe-based visual inertial slam nonlinear optimization [3], gas networks, beam lines, packing ellipsoids optimization and deep excavation analysis [4], as well as compressed sensing and parallel computing [5].

Except classical and common optimization methods, some new nonlinear optimization approaches and dynamic models are also proposed to solve the convex, facility layout, and limit analysis problems with fuzzy related constraints [6, 7]. Despite that the constraints are usually diverse in different optimization problems where continuously differentiable objective functions with bound and inequality constraints under uncertainty and binary variables are making the optimization problems more complicated [8], they can be well solved by using a multiparametric mixed-integer and sparse-based algorithm as well as linear programming, recurrent neural network, Bayesian optimization library, and direction interior-point techniques [9]. The result manifests that these methods can not only build a framework for dynamic optimization problems but also identify better integer solutions with promising feasibility [10]. Other strategies including the machine learning algorithms for hyperparameter optimization [11], pattern and stochastic search [12, 13] are also becoming a new trend in the combined optimization models, which are also demonstrated to be valid in solving parameter optimization and real-time nonlinear optimization problems.

Performance evaluation methods regarding nonlinear optimization are also investigated in related reports. Except the basic performance indicators such as the hypervolume, running time, and contribution, a large amount of evaluation methods and techniques are reported [14, 15]. Aiming at the robustness and convergence analysis, an approximation technique is presented [16], which employs the respective dual norm of the first-order derivatives of constraints to form two sparsity preserving ways for efficient computation of the derivative in solving large-scale and control problems [17]. As for the complexity and model structural stability analysis [18], various matrix conditioning and numerical analysis methods within a unifying framework of a proximity-scaling or a threshold technique [19] are applied, which have been presented as an effective example in process engineering. Nevertheless, human activities and related constraint conditions frequently affect the nonlinear optimization process. Solving nonlinear optimization problems more efficiently and with better models is still a challenge under limited resources, especially when large-scale and high-dimensional optimization applications are encountered in practice.

It is worth noting that the particle swarm optimization (PSO) method is one of the most important meta-heuristic optimization methods [1, 17] and has gained increasing attention in nonlinear problem solving, which can achieve the goal of optimization by simulating the behavior of particle swarms and searching for the best solution through updating the particle moving displacement and velocity. When an optimal PSO model is built, it can be applied in various fields and solve a variety of practical problems. For instance, the optimal design of the corona ring on high voltage composite insulators [20], optimal free-defect synthesis of four-bar linkage with joint clearance [21], and fast nonlinear predictive control on field programmable gate array (FPGA) can be achieved by PSO with dynamic population size [22]. Theoretically, constrained multivariable predictive controllers with model uncertainty and portfolio optimization problem [23], parameter optimization of nonlinear gray Bernoulli model [24], nonlinear regression, and unconstrained nonlinear optimization problems can also be well solved efficiently. However, the optimization efficiency and accuracy are usually influenced and limited by specific application situations and various constraints.

Consequently, diverse improved PSO models are proposed to alleviate the challenges mentioned above. For instance, to solve the mixed integer nonlinear optimization problem where the practical hydrothermal scheduling system with an inequality constraint treatment mechanism is highly complex, an improved PSO-based discrete binary and dynamic search-space squeezing strategy is proposed [25], the numerical calculation results compared with the traditional dynamic, nonlinear, and evolutionary programming methods show that the improved PSO behaves the best in terms of convergence speed, running time, and minimum cost. To identify the nonlinear system faster, more accurate, and choose the inertial weights and constriction coefficients more properly, modified PSO models for locating better local minimum are also proposed in [26]. Chaotic PSO [27, 28] and the enhanced PSO using exponential probability distribution testing results manifest their effectiveness in nonlinear optimization performance improvement [29, 30].

In recent years, aiming at specific applications and theoretical problems, numerous nonlinear optimization methods are proposed. A mixed logarithmic barrier-augmented Lagrangian method and GA algorithm for solving nonlinear optimization problems defined with fuzzy relational equations and max-Lukasiewicz composition are presented in [31], where the excellent computational performance is highlighted. Furthermore, a novel five-step discrete-time forms of Zhang dynamics (DTZD) and state-space decomposition algorithm [32] for time-varying nonlinear optimization, an approximation scheme [33] for distributionally robust nonlinear optimization, and a user preference enabling method [34] for constrained multi-objective nonlinear optimization are all reported. As for PSO-related nonlinear optimization methods, they are mainly focused on the combination of meta-heuristic optimization technologies, time-varying and nonlinear control system prediction. Such as the hybrid methods [35] are proposed to solve the nonlinear optimization problem evaluated by the reliability of a complex system, identification accuracy, and convergence speed, which have been efficiently used in various engineering fields, for instance, the parameter recognition for nonlinear dynamic hysteretic models [36] and industrial robot [37], dynamic wireless sensor network optimization [38] in data transfer acceleration and energy losses reduction, as well as the optimization of electromagnetic geophysical data [39]. Although some improved PSO models with Brownian Motion are reported [40, 41], they are usually separately applied in the optimization process, where the evolutionary mechanisms of PSO are unchanged. Such as, the Fractional Brownian Motion (FBM) model was proposed in [40] to forecast a non-stationary time series, where the Quantum-Behaved PSO (QPSO) is only utilized to optimize the multifractional hurst exponent, while the PSO and FBM are independently considered. In addition, the machine learning-based algorithms of support vector machine (SVM) and PSO combined algorithms are gradually considered in the precipitation data prediction [42], which is also the future research direction of nonlinear optimization. In other words, although tremendous advances in nonlinear optimization algorithms and software have been seen in recent decades, combining modern intelligent algorithms and platforms is the key to solving big data or large-scale problems with real-time strategies.

Motivated by the Euler-Maruyama (EM) principles and the optimization process of the PSO model, a nonlinear optimization method based on improved PSO algorithm is proposed here.

2. Models and algorithms

In this section, the basic theories of the improved PSO-based optimization models and principles are introduced in detail. The frameworks of the improved PSO and nonlinear optimization algorithms are also described and explained.

2.1 Models and principles

The traditional Brownian particle motion in the moment of t can be described as a standard normal distribution, and the distribution probability density is defined as follows:

$$N(x,t) = \frac{1}{\sigma\sqrt{2\pi t}} e^{-\frac{x^2}{2\sigma^2 t}} \quad (1)$$

However, the random motion mode of these particle swarms is usually related to the searching efficiency, accuracy, convergence performance, and solving speed. Consequently, the Euler-Maruyama (EM) principle is considered to solve the particle motion equation, namely,

$$x(t) = x_0 + \int_0^t f(x(s))ds + \int_0^t g(x(s))dw(s) \quad (0 \leq t \leq T) \quad (2)$$

where $x(t)$ is the solution of a stochastic differential equation (SDE), x_0 is the initial value of $x(t_0)$, f and g are the scalar functions, and w is a scalar standard Brownian motion within the domain of [0, T]. The differential form of (2) can be calculated by

$$dx(t) = f(x(t))dt + g(x(t))dw(t) \quad (x(0) = x_0, 0 \leq t \leq T) \quad (3)$$

Let $\Delta t = T/L, \tau_i = i\Delta t$, where L is the number of the time interval, which is a positive integer, and i is the evolution or iteration sequence. Thus, the SDE can be obtained as a discretized form of

$$x_i = x_{i-1} + f(x_{i-1})\Delta t + g(x_{i-1})\Delta w(t) \quad (4)$$

where the increment of $\Delta w(t)$ is defined as

$$\Delta w(t) = w(\tau_i) - w(\tau_{i-1}) \quad i = 1, 2, \ldots, L \quad (5)$$

The integral form of (4) can be computed and expressed as

$$x(\tau_i) = x(\tau_{i-1}) + \int_{\tau_{i-1}}^{\tau_i} f(x(s))ds + \int_{\tau_{i-1}}^{\tau_i} g(x(s))dw(s) \quad (6)$$

Assuming that the scalar functions of $f(x)$ and $g(x)$ are expressed as the approximate forms of $f(x) = \lambda x$ and $g(x) = \mu x$, the differential forms of (6) can be obtained as follows:

$$dx(t) = \lambda x(t)dt + \mu x(t)dw(t) \quad (7)$$

where λ and μ are constants, and the solution of (7) is calculated as follows:

$$x(t) = x_0 \cdot e^{\left(\left(\lambda - \frac{\mu^2}{2}\right)t + \mu w(t)\right)} \tag{8}$$

If the time interval of Δt can be divided into a smaller form of δt by a new introduced constant of R, namely,

$$R = \frac{\Delta \mathrm{t}}{\delta t} \tag{9}$$

the increment of $w(t)$ can be further expressed as a cumulative form of

$$\Delta w(t) = w(\tau_i) - w(\tau_{i-1}) = \sum_{k=iR-R+1}^{iR} dw_k \tag{10}$$

Combining the traditional PSO algorithms, the displacement and velocity of the moving particle of X_i and V_i can be depicted as

$$X_\mathrm{i} = X_i + V_i \tag{11}$$

$$V_i = V_i + c_1 r_1 \cdot (P_{best} - X_i) + c_2 r_2 \cdot (G_{best} - X_i) \tag{12}$$

where c_1 and c_2 are the constants, r_1 and r_2 are the random numbers, and P_{best} and G_{best} are the local best solution and global best solution in the optimization and iteration process. However, the optimization process is usually influenced by the random number of r_1 and r_2 greatly. If they are replaced by the EM solution of $x(t)$ expressed in (8) when (9) and (10) are considered, an improved PSO model can be obtained and expressed as follows:

$$\mathrm{r}_1 = r_2 = x_0 \cdot e^{\left(\left(\lambda - \frac{\mu^2}{2}\right)t + \mu w(t)\right)} \tag{13}$$

It can be found that the random optimization process within the lower and upper searching space are improved as a discretized and more efficient way, which is expected to enhance the optimization efficiency and accuracy by replacing the random optimization process with a more regular way when moving and updating these particle swarms.

2.2 Algorithms

Specifically, the framework of the improved PSO algorithm is expressed in Algorithm 1 as follows:

Algorithm 1: Improved PSO method.

---

1 Initialize the basic parameters of the particle population including the moving velocity V_i and position X_i, lower and upper bound on search space of X_{low} and X_{upper}, population number N_p, the maximum iteration number N, and set the initial best position as $P_{best} = X_i$;

2 Evaluate the moving state of each particle and set the global best position of $G_{best} = \min\{P_{best}(1), P_{best}(2), P_{best}(3), \ldots P_{best}(\mathrm{i})\}$;

3 **While** G_{best} is not found
4 **For** i = 1 to N
5 Calculate the particle moving state parameters of r_1 and r_2 according to (13);
6 Calculate the fitness value of each particle for the moving velocity and position;
7 **If** the fitness value of $F(X_i) < F(P_{best})$
8 Update the position with $P_{best} = X_i$;
9 **If** the fitness value of $F(P_{best}) < F(G_{best})$
10 Update the global best position with $G_{best} = P_{best}$;
11 **End if**
12 Update the moving velocity of V_i according to (12);
13 **End for**
14 **End while**
15 Obtain the final result of global best value G_{best};
16 **End**

Explanatorily, the moving position and speed of the particle swarm are first initialized randomly within the lower and upper bound. Then, the fitness value of each particle is calculated until the maximum iteration number is reached. Further, the position and velocity of the particles are compared and updated if the local best solution is searched, and the global best solution is consequently obtained by comparing the best local solutions in each iteration. Thereafter, the particle moving position and velocity can be updated combining (12) and (13). Finally, the global best value of the objective function can be obtained in the PSO when the termination condition is satisfied, and the circulative iteration is finished. When the improved PSO method is applied to solve nonlinear optimization problems, the framework is presented as the process of Algorithm 2.

Algorithm 2. Nonlinear optimization process by improved PSO method

---

1 Set up the nonlinear problem with specific models;
2 Initialize the basic parameters in the nonlinear model;
3 Initialize the basic parameters of PSO models, including the maximum iteration number, lower and upper bounds of the search space, initial value of variables in the nonlinear problem;
4 **While** N < maximum iteration number
5 Initialize population and particle swarm;
6 Search for the global best nonlinear value at each iteration of particle evolution;
7 Update particle moving speeds and positions with (11) and (12) through the improved PSO;
8 Update the search direction with the optimal hyperparameters;
9 Evaluate the nonlinear model and compute the global best values;
10 Update the best solution for every nonlinear problem;
11 Obtain the nonlinear optimization result
12 **End while**

First, the constrained or unconstrained nonlinear problems should be modeled in a mathematic form. Then, the basic parameters of variables in the nonlinear model are initialized with starting values, as well as evolution parameters in the improved PSO, such as $\lambda = 0.1$, $\mu = 1$, $R = 4$, population = 300, $N = 30$, $C_1 = 0.5$, $C_2 = 0.5$, and the weight of ω = 0.7. Then, the nonlinear optimization is conducted within a circulative iteration, where the global best nonlinear values can be obtained after updating the particle moving positions and speeds in each of the evolution step. Consequently, the search direction with optimal hyperparameter is updated, and the nonlinear optimization result is evaluated until the best solution is calculated. Consequently, the new enhanced PSO algorithm can be utilized in nonlinear parameter optimization by improving the particle evolution and searching forms with better convergence performance.

3. Optimization characterization

In this section, the nonlinear optimization characterization is mainly investigated from the perspective of model parameter and dimensional influence. Besides, the nonlinear optimization method based on the improved PSO algorithm is preliminarily validated by different nonlinear optimization problems.

3.1 Influence of model parameters

When the improved PSO models are built, their optimization performance is usually influenced by model parameters including the constants of λ, μ, and R mentioned in (7) and (9), as well as the population size of the particle swarm, iteration number of N, C_1, and C_2. Typically, these parameters are set as λ = 0.1, μ = 1, R = 4, population = 300, N = 30, C_1 = 0.5, C_2 = 0.5, and the weight of ω = 0.7, when one of them is changed according to the parameter setting expressed in **Table 1**, other parameters remain unchanged. There are totally nine groups of variables for each of the optimization model parameter.

On the other hand, a typical nonlinear optimization problem based on Bohachevsky test function is mainly studied, and the objective function is defined as,

$$o(x) = {x_1}^2 + 2{x_2}^2 - 0.3\cos(3\pi x_1) - 0.4\cos(4\pi x_2) + 0.7 \tag{14}$$

This nonlinear function is investigated within the domain of $x_i \in [-100, 100]$ for all i = 1, 2, and there are two variables or dimensions in the objective function. When the global minimum of the nonlinear function is searched by the proposed improved PSO

Serial no.	**1**	**2**	**3**	**4**	**5**	**6**	**7**	**8**	**9**
λ	0.01	0.05	0.1	0.2	0.5	1	2	5	10
μ	0.1	0.2	0.5	1	2	5	10	20	50
R	1	2	3	4	5	7	10	20	50
Population	50	100	200	300	400	500	600	800	1000
N	10	20	30	50	60	80	100	200	500
ω	0.01	0.1	0.2	0.3	0.5	0.6	0.7	0.8	1
C_1	0.01	0.05	0.1	0.2	0.3	0.5	0.6	0.8	1
C_2	0.01	0.05	0.1	0.2	0.3	0.5	0.6	0.8	1

Table 1.
Parameter setting in different comparing models.

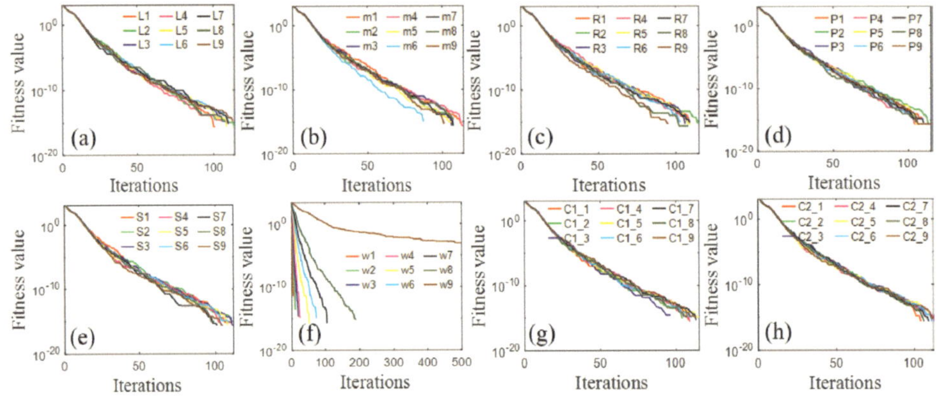

Figure 1.
Influence of eight model parameters to the algorithm convergence: (a) λ*; (b)* μ*; (c)* R*; (d) population size; (e)* N*; (f)* ω*; (g)* C_1*; and (h)* C_2*.*

based method, the convergence performance characterized by the fitness values in each of the iteration under different model parameters is illustrated and expressed in **Figure 1**. Explanatorily, the legends of *L1* to *L2* appeared in **Figure 1(a)** represent the parameter values of λ changed from 0.01 to 10 according to the second line mentioned in **Table 1**. Likewise, when the improved PSO model parameters of μ, R, population, N, ω, C_1 and C_2 change, the fitness values in each of the iterations and optimization process are calculated and presented in **Figure 1(b)-(h)**, respectively.

Obviously, all these fitness values are decreased gradually in the evolution process no matter how these parameters change. Moreover, the convergence speeds are almost consistent when the model coefficients of λ, μ, R, population, N, C_1 and C_2 change observing from **Figure 1(a)-(e)** and **Figure 1(g)** and **(h)**. In other words, these seven parameters are not very sensitive to the convergence of the improved PSO-based nonlinear optimization algorithm. However, the convergence speeds and optimization performance of the proposed method are greatly influenced by the parameter of the searching weight ω, which can be found from the computing results presented in **Figure 1(f)**. It is evident that, as ω gains from *w1* to *w9* (namely, from 0.01 to 1), the convergence speed gradually decreases, and the maximum iteration number to search for the best global solutions increases progressively.

3.2 Influence of input dimension

To study the influence of the input dimension of the variable on the solving process and the final solution results, four typical nonlinear and multi-objective optimization functions from dtlz test suite including dtlz1a, dtlz2a, dtlz4a and dtlz7a [43] are chosen and investigated. There are two or three objective functions in the nonlinear optimization for each of the test problem, and the input dimensions are set as 2, 4, 6, 8, and 10 in the five control groups. After calculating for these four dtlz test functions, five groups of Pareto set (PS) for each of the multi-objective and nonlinear problems can be obtained, respectively, as shown in **Figure 2**.

When the input dimension of d increases from 2 to 10, the PS calculation results for dtlz1a test problem keep nearly the consistent, which means that the nonlinear optimization and evolution process are immune to the change of the input dimension,

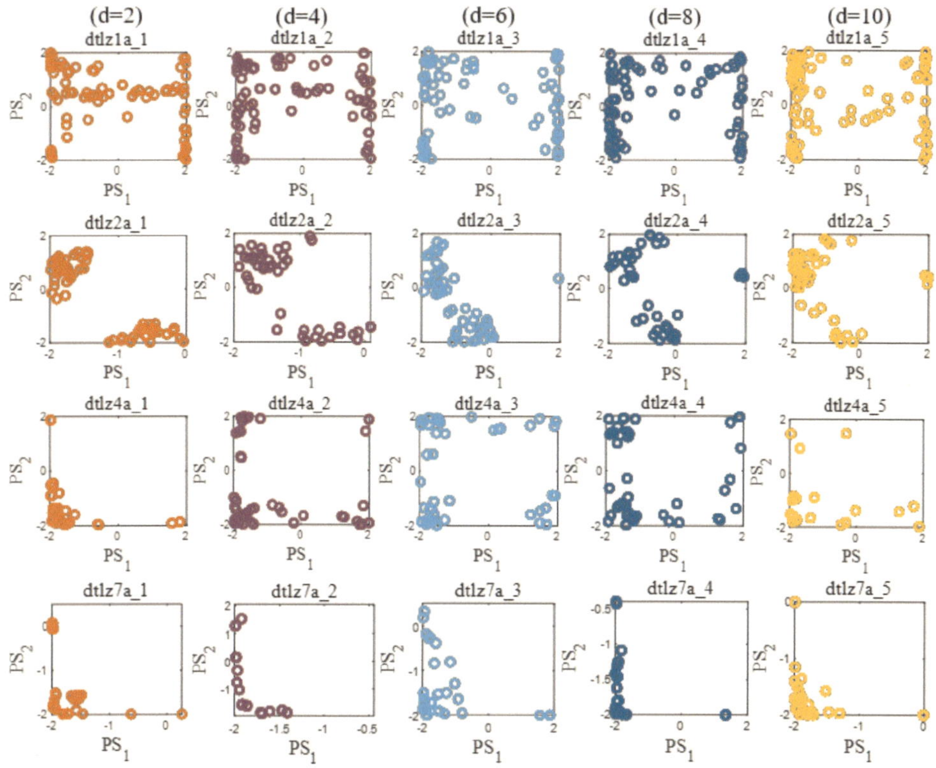

Figure 2.
Influence of the input dimension to Pareto set.

and the improved PSO-based nonlinear optimization method is robust and reliable in solving dtlz problems. Besides, the PS computing results acquired for the other dtlz problems of dtlz2a, dtlz4a, and dtlz7a also indicate that the optimal solutions can be well found when the input dimension gains from 2 to 10, and the distribution of these PS results are almost consistent. On all accounts, the proposed nonlinear algorithm can not only solve the dtlz problems but also can maintain a relative higher robustness and reliability when the input dimension changes.

3.3 Study on different nonlinear problems

What's more, eight different nonlinear problems are further investigated and studied with various output variables including the MMF1 to MMF3, kno1, oka1, oka2, vlmop2, and vlmop3 functions [44]. When the improved PSO-based nonlinear optimization method is applied to solve these problems, the calculation results of the Pareto set are obtained and expressed in **Figure 3**.

Generally, these nonlinear models have two main objective functions except vlmop3. Although the eight nonlinear problems are different in mathematic definition and global characterization, all their Pareto set can be acquired exactly. Namely, the optimization method is capable of solving various nonlinear problems especially for multi-objective models, while the robustness and reliability can always keep consistent regardless of the change of model parameters. Besides, the preliminary

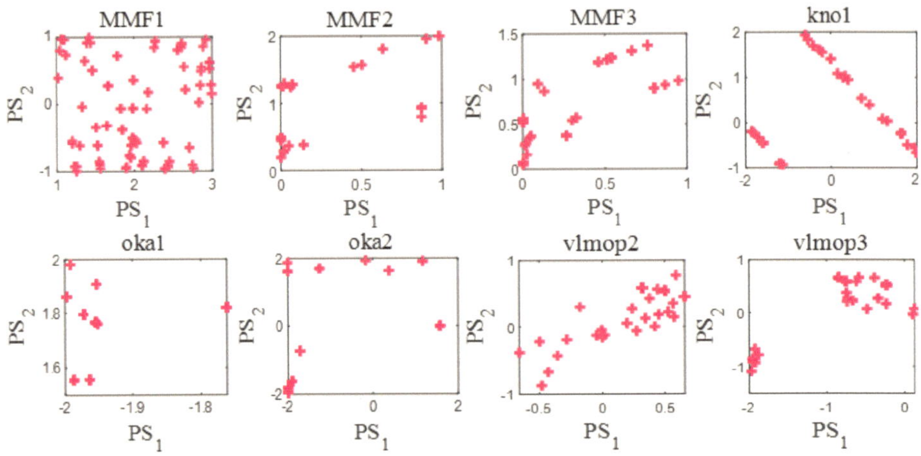

Figure 3.
Influence of different nonlinear problems on the new PSO calculating results.

calculation results of Pareto set through different nonlinear models also demonstrate its feasibility in nonlinear optimization and sophisticated problem solving.

4. Nonlinear optimization results

Based on the model parameter investigation and nonlinear optimization performance characterization, more nonlinear optimization results including the Pareto fronts (PFs) obtained from different nonlinear optimization methods and problems are studied and compared in this section. The advantages of the proposed improved PSO based method are further verified.

4.1 Comparison of Pareto fronts

First, the PFs computed for these 12 different nonlinear models including MMF1 to MMFL3, vlmop2, vlmop3, oka1, oka2, kno1, dtlz1a to dtlz7a as mentioned above are presented, where 12 groups of PFs calculation results by the new PSO-based nonlinear optimization algorithm are also presented and compared in **Figure 4**. Observed by the results from PF1 to PF12, it can also be found that almost all the concavity and convexity for these nonlinear models are well exhibited even for MMF2 and oka2. Namely, better nonlinear optimization results of PFs can be obtained from the new PSO-based algorithm with more efficient particle moving patterns relative to the traditional PSO method with random way.

4.2 Comparison of different methods

According to the principle of steepest decent algorithm (SDA), if the optimized nonlinear function of $f(X)$ has first-order continuous partial derivative, after the iteration for k times, the searching direction for the global best solution is defined as follows:

$$S^k = -\nabla f\left(X^k\right) \tag{15}$$

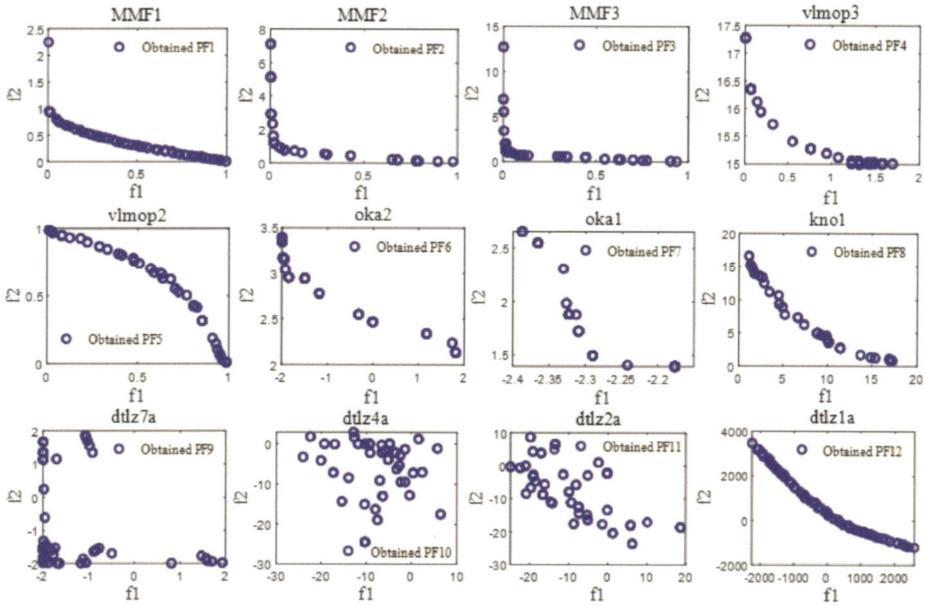

Figure 4.
PFs calculated from the new improved PSO model.

However, the convergence speed of the traditional SDA is slow, and the saw-tooth phenomenon will appear especially when the searching space is near the minimum value. As for the conjugate gradient algorithm (CGA), when an initial point of X^0 and a descent direction of S^0 is set, the first searching result of X^1 can be obtained as

$$\nabla f(X^1)^T S^0 = 0 \tag{16}$$

To overcome the saw-tooth phenomenon occurred in the steepest decent method, twice searching for a binary quadratic nonlinear equations is needed to find the minimum value. Then, a conjugate search direction vector for S^0 which satisfies the following equation is introduced:

$$(S^0)^T A S^1 = 0 \tag{17}$$

where A is a symmetric positive matrix and satisfies

$$f(X) = \frac{1}{2} X^T A X + b^T X + c \quad (\mathrm{b} \in E^n, \quad \text{c is a constant}) \tag{18}$$

Consequently, the conjugate search direction of S^1 is obtained as follows:

$$S^1 = -\nabla f(X^1) + \frac{(S^0)^T A \nabla f(X^1)}{(S^0)^T A S^0} S^0 \tag{19}$$

if a binary quadratic nonlinear function is searched along S^0 with a random initial point of X^0 and then searched through the conjugate direction of S^1, the minimum solution of X^* can be acquired through CGA.

Besides, another classical variable metric method called Davidon-Fletcher-Powell algorithm (DFPA) is also applied and compared. When a scale matrix of H_k^{-1} is considered and calculated as follows:

$$H_{k+1} = H_k + \frac{\Delta X^k \left(\Delta X^k\right)^T}{\left(\Delta X^k\right)^T \Delta g_k} - \frac{H_k \Delta g_k \cdot \Delta g_k{}^T H_k}{\Delta g_k{}^T H_k \Delta g_k} \tag{20}$$

where g_k and X_k are described as

$$\begin{cases} g_k = \nabla f\left(X^k\right) \\ \Delta g_k = g_{k+1} - g_k \\ \Delta X^k = X^{k+1} - X^k \end{cases} \tag{21}$$

Then, the searching direction of S_k can be obtained as

$$S_k = -H_k g_k \tag{22}$$

DFPA is efficient in nonlinear optimization when searching for the global minimum, but the round-off error and imprecise searching may lead to the existence of singularity for H_k. Thus, an improved method of Broyden-Fletcher-Goldfarb-Shanno algorithm (BFGSA) is proposed, where the positive definite matrix of H_k satisfies,

$$H_{k+1} = \frac{H_k + \left[\left(1 + \frac{\Delta g_k{}^T H_k \Delta g_k}{\left(\Delta X^k\right)^T \Delta g_k}\right) \Delta X^k \left(\Delta X^k\right)^T - H_k \Delta g_k \left(\Delta X^k\right)^T - \Delta X^k \Delta g_k{}^T H_k\right]}{\left(\Delta X^k\right)^T \Delta g_k} \tag{23}$$

Although BFGSA has similar mathematical features to that of the DFPA, the singular matrix will not be easily generated, and it has more stable numerical solutions and requires less searching precision than DFPA.

Furthermore, five groups of typical nonlinear optimization test functions are applied and tested, where the 2D sphere function (T1), Himmelblau's function (T2), Goldstein-Price function (T3), 2D Styblinski-Tang function (T4), and 5D Styblinski-Tang function (T5) are included, and the definition of these test problems is as follows:

$$\begin{cases} T1 = \sum_{i=1}^{n} x_i{}^2 \\ T2 = \left(x_1{}^2 + x_2 - 11\right)^2 + \left(x_1 + x_2{}^2 - 7\right)^2 \\ T3 = \left[1 + \left(x_1 + x_2 + 1\right)^2 \cdot \left(19 - 14x_1 + 3x_1{}^2 - 14x_2 + 6x_1x_2 + 3x_2{}^2\right)\right] \\ \cdot \left[30 + \left(2x_1 - 3x_2\right)^2 \left(18 - 32x_1 + 12x_1{}^2 + 48x_2 - 36x_1x_2 + 27x_2{}^2\right)\right] \\ T4 = \frac{1}{2} \sum_{i=1}^{2} \left(x_i{}^4 - 16x_i{}^2 + 5x_i\right) \\ T5 = \frac{1}{2} \sum_{i=1}^{5} \left(x_i{}^4 - 16x_i{}^2 + 5x_i\right) \end{cases} \tag{24}$$

When these classical nonlinear optimization methods are applied to these functions test, the global best values and the optimized objective values, as well as the iteration numbers are acquired and shown in **Table 2**. Although the global best values for the test functions of T1, T2, T4, and T5 are accurately calculated with few iterations, there exist relative great optimization errors for the optimization results of T3.

Besides, the similar PSO algorithms expanded as PSO_2 and PSO_3 are also compared. Different with the traditional PSO algorithm (PSO_1), the PSO_2 is modeled with a cumulative form of r_i (i = 1, 2), for instance,

$$r_1 = r_2 = \sum \sqrt{\frac{1}{N}} \cdot rand(N) \tag{25}$$

Likewise, the expanded model of PSO_3 referring to the reasoning process of the improved PSO is defined as

$$r_1 = r_2 = x_0 \cdot e^{\left(\left(\lambda - \frac{\mu^2}{2}\right)t + \mu \cdot \sum \sqrt{\frac{1}{N}} \cdot rand(N)\right)} \tag{26}$$

Except the random term of r_i, the extended models of PSO_2 and PSO_3 have the same optimization and evolution process with that of the improved PSO as mentioned before in Algorithm 2. After calculating for the test functions of T1 to T5 by the comparative PSO_1 to PSO_3 and the proposed improved PSO method, the nonlinear optimization results are obtained and presented in **Table 3**.

Comparing the optimization results shown in **Tables 3** and **4**, the global minimum values of T1 to T3 can all be obtained exactly through these four meta-heuristic algorithms of PSO, especially for T3, while the classical four algorithms mentioned in **Table 3** fail to solve. Besides, T4 can also be solved well by PSO_1, PSO_3, and the improved PSO method. As for T5, PSO_1 and the improved PSO can all be optimized with relative smaller errors compared with the true global best values. In other words, the improved PSO method behaves the best among these four PSO models in the nonlinear optimization, which is even better than that of these four classical optimization methods mentioned in **Table 3**.

4.3 Comparison of different nonlinear problems

Different nonlinear problems are further studied in this section to validate the feasibility and reliability of the proposed improved PSO method. Typically, 12 nonlinear models including F1 to F12 are mainly investigated and tested. The butterfly optimization algorithm (BOA), the traditional PSO method, hybrid PSO with BOA (HPSOBOA), as well as the new improved PSO model (PSO_new) are applied and compared in the F series functions test. After the calculation and evolution in the nonlinear problem testing, the change trend of fitness values in the global minimum and best solution searching or iteration process are obtained, as expressed in **Figure 5**. It can be clearly seen from the fitness calculation results for F1, F2, F3, F4, F5 and F7, F9, F10, F12 that the new proposed PSO method has the fastest convergence speed and least iteration numbers in searching for the global best solution of these nine nonlinear problems.

As for the left F6, F8, and F11, the new PSO also behaves slightly better than that of the traditional PSO method. In brief, the new proposed PSO method not only has better nonlinear optimization performance than that of the BOA and HPSOBOA in

Function	Global Best value	SDA		CGA		DFPA		BFGSA	
		Iteration	Objective value	Iteration	Objective value	Iteration	Objective value	Iteration	Objective value
T1	0	1	0	2	0	1	0	1	0
T2	0	8	0	7	0	5	0	5	0
T3	**3**	22	**30**	7	**30**	5	**30**	5	**30**
T4	−78.33	1	−78.3323	2	−78.3323	1	−78.3323	1	−78.3323
T5	−195.83	1	−195.8308	2	−195.8308	1	−195.8308	1	−195.8308

Table 2.
Nonlinear optimization results from four classical methods.

Function	Global Best value	PSO_1		PSO_2		PSO_3		Improved PSO	
		Iteration	Objective value	Iteration	Objective value	Iteration	Objective value	Iteration	Objective value
T1	0	20	0	20	0.000152	20	0	9	0
T2	0	50	0	50	0	50	0	46	0
T3	3	50	3	50	3	50	3	39	3
T4	−78.33	50	−78.30	50	−78.20	50	−78.30	37	−78.30
T5	−195.83	25	−195.8257	25	−195.6255	25	−195.00	25	−195.8153

Table 3.
Nonlinear optimization results from different PSO models.

most cases but also features faster convergence speed and higher efficiency in searching for the global best solutions than that of the traditional PSO. Consequently, the feasibility and reliability of the improved PSO is validated, which is expected to be applied to more sophisticated nonlinear problem solving and optimization in practical applications.

5. Performance evaluation and application

In this section, the proposed improved PSO based nonlinear optimization method is mainly evaluated by the numerical experiments and performance indicators. Besides, comparisons regarding different PSO models and nonlinear problems are also conducted and investigated. Finally, a nonlinear parameter optimization case study, typical CEC benchmark problems, and rank sum test are also considered, which all demonstrate the feasibility and preferable performance of the proposed improved PSO method.

5.1 Experimental setup

The numerical experiments are mainly conducted through the platform of MATLAB R2022b with the computer configuration of Intel(R) Core (TM), i7-7700HQ CPU @ 3.2GHz and 16.0GB RAM, 64-bit Windows operating system. Besides, more than 12 nonlinear optimization problems including the MMF functions, vlmop, and dtlz series functions as mentioned above are tested. Four related PSO models are mainly applied in the nonlinear solution comparison, where the main parameters are set as $\lambda = 0.1$, $\mu = 1$, $R = 4$, population = 300, $N = 30$, $C_1 = 0.5$, $C_2 = 0.5$, and the weight of $\omega = 0.7$. Furthermore, all the numerical experiments are conducted for 50 times to guarantee repeatability and reliability.

5.2 Hypervolume and time

The hypervolume indicator is primarily used to measure the performance of the nonlinear optimization results, which is defined as follows [45]:

$$Hv(S) = \Lambda(\{q \in \mathbb{R}^d | \exists p \in S : p \leq q, q \leq r\}) \tag{27}$$

where S is the given point set and r is a reference point, and S is the region measure weekly dominated by the bound of r, and Λ is the Lebesgue measure. The traditional PSO method and the improved PSO models with different parameters are applied in the tests for 12 nonlinear problems. The optimization indicators of hypervolume (Hv) and running time (/s) of these algorithms calculated in each of the model are obtained and presented in **Table 4**, where the population number of the particle swarm and the coefficient of λ are annotated in the brackets.

It can be seen that all the parameters of λ are set as 0.1, while the population for the traditional PSO is 400 and diverse for the improved PSO. As the population increases from 200 to 400, the hypervolumes are also increased slightly in each of the test problems through the improved PSO method, while the running time changes a little. Namely, the robustness of the improved PSO method is increased while the optimization efficiency is unaffected as the population size gains. Noteworthily, when the

Problems	PSO (400, 0.1)		Improved PSO (200, 0.1)		Improved PSO (300, 0.1)		Improved PSO (400, 0.1)	
	Hv	Time	Hv	Time	Hv	Time	Hv	Time
MMF1	3.6438	2.7727	3.6372	2.9084	3.6274	2.7539	3.6401	2.6624
MMF2	3.4524	3.7696	3.5099	4.1256	3.4987	3.9866	3.5277	4.1519
MMF3	3.4532	4.4021	3.4052	3.9400	3.5574	3.8329	3.5124	4.0124
vlmop3	−26.663	3.4645	−26.678	3.1630	−26.8320	3.1953	−26.7590	3.5025
vlmop2	3.3231	2.9952	3.2624	3.7602	3.2790	4.2806	3.2945	3.8899
oka2	−2.2072	3.6441	−2.1605	3.8799	−2.4836	3.4863	−2.5338	3.3566
oka1	3.2775	8.4793	1.7238	7.3141	2.6560	8.1473	3.5655	8.2796
kno1	−61.9418	4.2376	−73.6724	5.1355	−91.1023	4.2069	−93.3501	4.6448
dtlz7a	14.6180	2.7965	12.5099	2.3893	15.8068	3.1955	14.6047	2.6112
dtlz4a	208.0310	2.9061	167.9735	2.3034	442.6884	2.2043	331.8512	2.3637
dtlz2a	397.6778	2.9043	41.0199	2.2796	355.5462	2.6709	292.8354	2.7049
dtlz1a	−4.211e6	1.3554	−6.917e6	1.16905	−7.558e6	1.1458	−8.836e6	1.2760

Table 4.
Hypervolume and running time calculation results.

population and λ are set as 300 and 0.1, most of the hypervolume is bigger than that of the traditional PSO, and the vast majority of running time is less than the traditional PSO. It manifests that the new proposed PSO is more capable of nonlinear optimization with higher calculation efficiency, stability, and robustness.

5.3 Comparison of nonlinear solutions

The optimization process is further studied in this section when four typical nonlinear test problems of T1 to T4 as defined and mentioned before in (24) are considered. Besides, four related PSO methods including the traditional particle swarm optimization method of PSO_1, expanded models of PSO_2 and PSO_3, as well as the new proposed improved PSO model of PSO_4 are applied and compared. The optimization and global searching process can be visualized and presented in **Figure 6**.

Observed by the nonlinear optimization process for T1, T3, and T4, it can be found that although the PSO_1 can search the accurate global best solutions, the searching space is limited and the particles usually gather within a small scope, which may lead to the leak of some local extreme points and important positions. As a contrast, the searching coverage from the improved PSO_4 is greatly extended, which makes it more efficient and accurate in finding the local best and global best values in the nonlinear optimization problems solving. However, relative bigger searching errors may be generated when the PSO_2 for T4 and PSO_3 for T2 are investigated. In a word, the feasibility of the improved PSO method with higher optimization efficiency and robustness is exhibited again from the perspective of optimization process observation and investigation.

5.4 Convergence analysis

The convergence of different PSO-based algorithms are also compared in this section; after the calculation and evolution for nonlinear test problems of T1 to T4, the solutions of objective values in each iteration are presented in **Figure** 7, where the legend of 'mean (Ti best)' (i = 1, 2, 3, 4) represents the mean of the best population value, while the legend of 'mean (Ti)' (i = 1, 2, 3, 4) denotes the mean of the current population value.

It can be found that the convergence curves for T1 to T3 by each of the corresponding PSO method are very alike, especially for the global best solution curves, which manifests that both PSO_2 and PSO_3 can solve the nonlinear optimization problems with a similar accuracy to PSO_1, as well as the improved PSO_4. Besides, no big differences are found among these PSO methods when observing the calculating results for T4, namely, all of them can search for the global best values within several iterations.

5.5 Wilcoxon rank sum test

Furthermore, the Wilcoxon rank sum test [46, 47] which recognizes the difference of two datasets is also considered to evaluate the benchmark problem-solving performance of these algorithms. Primarily, the comparison results indicated by binomial estimator, Hodges-Lehmann (HL) estimator, W value, Z value, mean, standard deviation (SD), as well as p value are calculated and presented in **Table** 5.

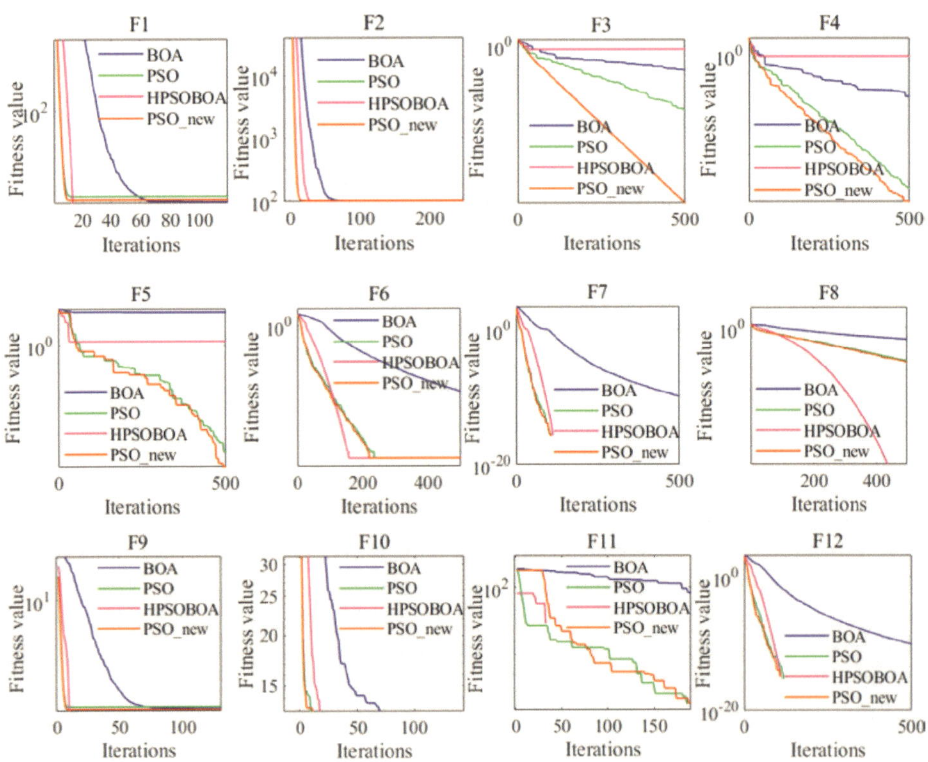

Figure 5.
Results of fitness for F functions.

Indicator	PSO_4 vs	For computing time					For minimum value				
		PSO	GA	OFA	CSO	SA	PSO	GA	OFA	CSO	SA
Binomial estimator	Mean differences	−0.0099	−0.0731	−0.072	−0.1244	−1.1323	−330	−18	−160	0.026	2300
	Confidence interval	[−0.0976, −0.0043]	[−0.1238, −0.0223]	[−0.1208, −0.0137]	[−0.3305, −0.0321]	[−2.6860, −0.5927]	[−7.00e10, −10.00]	[−1.00e10, 40.00]	[−2.40e11, 0.052]	[−3.70e11, 70.00]	[0.0180, 2.66e15]
HL estimator	Mean differences	−0.01905	−0.0733	−0.07165	−0.1658	−1.4183	−4.22e7	−1.15e6	−5.99e7	−9.785e7	7.05e8
	Confidence interval	[−0.0643, −0.0072]	[−0.1238, −0.0379]	[−0.1021, −0.0455]	[−0.2283, −0.0824]	[−2.7678, −0.8625]	[−7.6e15, −10.00]	[−1.5e11, 25.00]	[−1.5e15, −2.489]	[−3e15, 25.00]	[430, 5.16e16]
W		−448	−440	−426	−466	−528	−159	−93.0	−153	−79.0	205
Mean		0	0	0	0	0	0	0	0	0	0
SD		106.960	106.960	102.060	106.960	106.960	74.330	74.330	74.330	74.330	74.330
Z		4.1339	4.1092	4.1692	4.3522	4.9318	2.1324	1.2444	2.0517	1.0561	2.7512
p value		**1.433e-5**	**1.985e-5**	**1.529e-5**	**6.739e-6**	**4.073e-7**	**0.0165**	**0.1067**	**0.0201**	**0.1455**	**0.0030**

Table 5.
Comparison of Wilcoxon rank sum test by PSO_4 and other models.

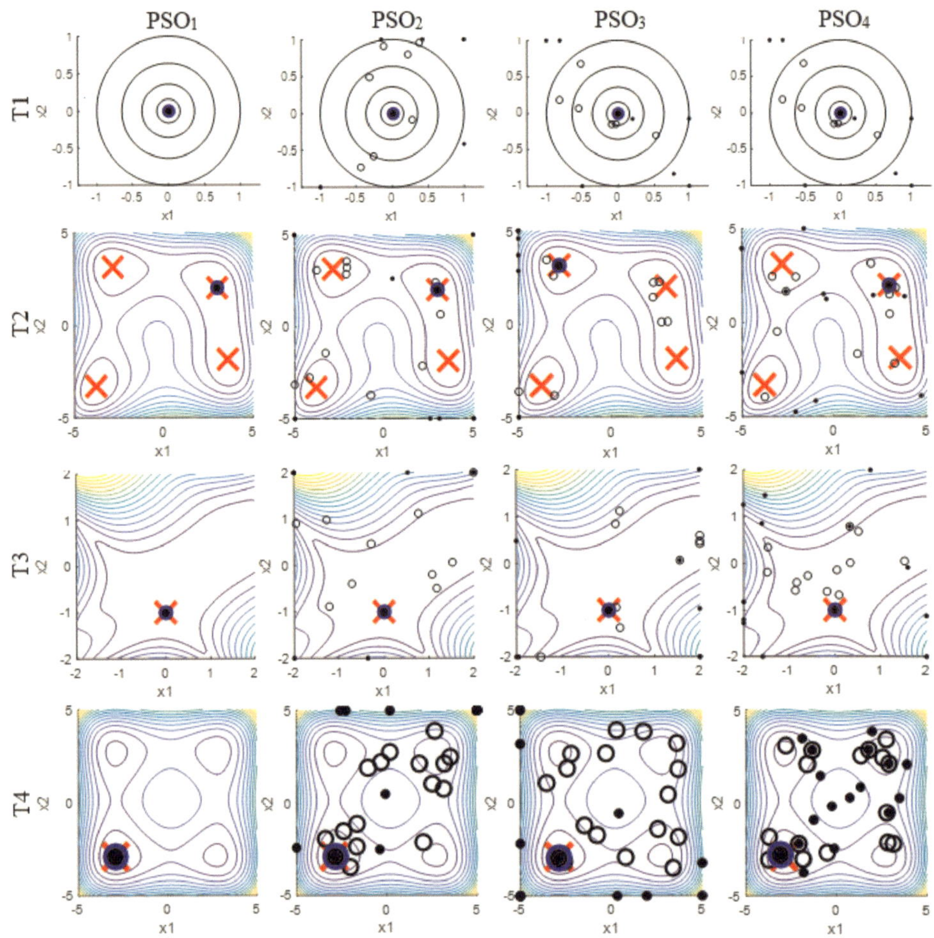

Figure 6.
Nonlinear optimization results by different test problems and PSO models.

These results show that there exists relative bigger difference between PSO_4 and PSO when calculating the running time, as well as the differences between PSO_4 and SA, while PSO_4 has low similarity in running speed performance and high similarity with these five typical algorithms in minimum value calculation and optimization observed by the indicators of mean differences and confidence interval. As for W, mean, SD and Z, all these models are alike. The p values for PSO, GA, OFA, CSO, and SA are all smaller than 0.05 in computing time, and they are also smaller than 0.05 for PSO, OFA, and SA, which manifests that the PSO_4 has superior performance in calculating speed or running time. Additionally, the new PSO_4 behaves more stable in solving CEC benchmark problems and has competitive performance in minimum value calculation and optimization.

6. Conclusion

This paper mainly proposes and investigates a nonlinear optimization method based on the improved particle swarm optimization algorithm. Different with the traditional

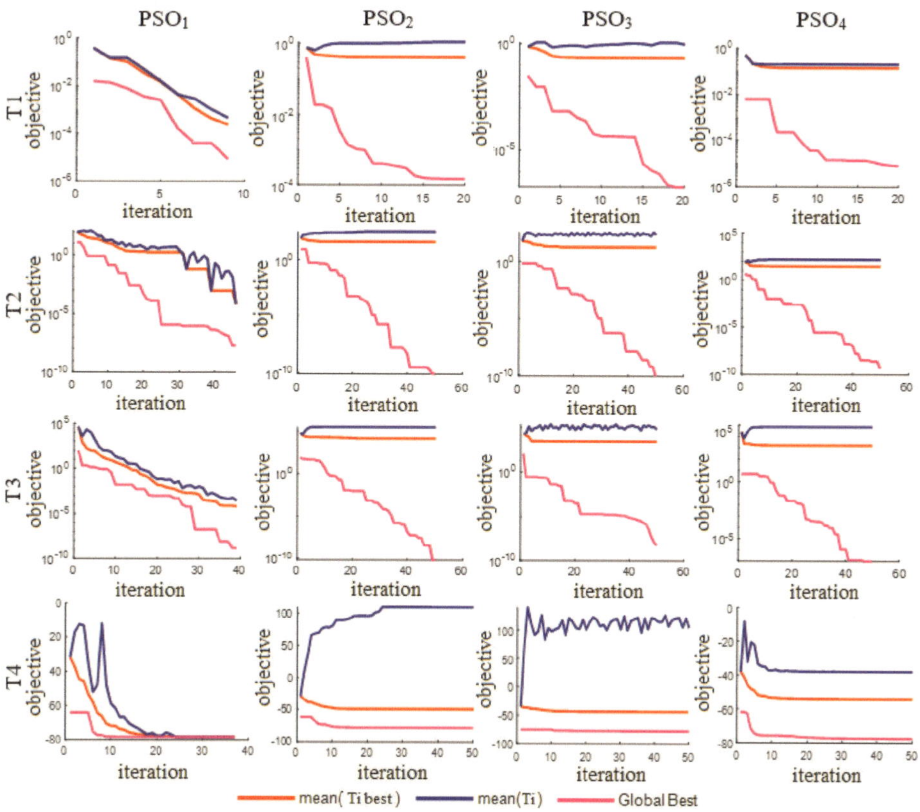

Figure 7.
Convergence curve for T1 to T4 test problems by PSO_1 to PSO_4 methods.

PSO method using a random searching pattern, the model and principles of the new PSO algorithms applying EM solving method are first introduced from the perspective of discretized solving process for SDE. Then, the influence of model parameters and input dimension on the nonlinear optimization characterizations are further revealed, and the feasibility of the proposed method is also validated regarding different nonlinear problems. Combining the optimization results of PFs and comparison with other nonlinear optimization methods, the capability and adaptability in solving various nonlinear problems are demonstrated and highlighted. Finally, the performance evaluation results manifests that the proposed method features faster convergence speed, accurate optimization results, as well as reliable and stable evolution process compared with other related optimization methods by CEC benchmark problem testing.

The future work may lie in solving high dimensional and more sophisticated nonlinear optimization problems with the adaptive parameter adjusting mechanism. Moreover, the promotion of the proposed nonlinear optimization method to various practical application is also of vital significance.

Acknowledgements

This work was financially supported by the National Natural Science Foundation of China (NSFC, Grant Nos. 52005203 and 52375542), Natural Science Foundation of

Hubei Province (Grant No. 2023AFB903), the Fundamental Research Funds for the Central Universities (Grant No. 2662022GXD002), and the foundation from National Key Laboratory of Electromagnetic Energy (Grant No. 61422172220507), and we would also like to acknowledge the valuable comments from the editor and reviewers.

Declaration of competing interest

The authors declare that they have no known competing financial interests that could have appeared to influence the work reported in this paper.

Data availability

The data used for the research will be made available on reasonable request.

Author details

Shiwei Liu[1]*, Yuandi Li[1], Longxiang Shan[1], Chengkang Liu[1], Dongqiao Wang[1], Qiaohua Wang[1], Xia Hua[2], Yanhua Sun[3] and Lingsong He[3]

1 College of Engineering, Huazhong Agricultural University, Wuhan, Hubei, China

2 National Key Laboratory of Electromagnetic Energy, Wuhan, Hubei, China

3 School of Mechanical Science and Engineering, Huazhong University of Science and Technology, Wuhan, China

*Address all correspondence to: hustliusw@sina.com

References

[1] Ruszczynski A. Nonlinear Optimization. Princeton: Princeton University Press; 2011

[2] Gould NI, Leyffer S. An introduction to algorithms for nonlinear optimization. Frontiers in Numerical Analysis. 2003: 109-197. Available form: https://link.springer.com/chapter/10.1007/978-3-642-55692-0_4

[3] Leutenegger S, Furgale P, Rabaud V, et al. Keyframe-based visual-inertial slam using nonlinear optimization. Proceedings of Robotis Science and Systems (RSS). 2013;**2013**: 37-44

[4] Tang Y-G, Kung GT-C. Application of nonlinear optimization technique to back analyses of deep excavation. Computers and Geotechnics. 2009; **36**(1–2):276-290

[5] Migdalas A, Toraldo G, Kumar V. Nonlinear optimization and parallel computing. Parallel Computing. 2003; **29**(4):375-391

[6] Chakraborty M, Kumar J. Lower bound limit analysis using nonlinear optimization for solving axisymmetric problems using Hoek-Brown yield criterion. International Journal of Geotechnical Engineering. 2021;**15**(1): 28-39

[7] Lu J, Fang S-C. Solving nonlinear optimization problems with fuzzy relation equation constraints. Fuzzy Sets and Systems. 2001;**119**(1): 1-20

[8] Murray W, Ng K-M. An algorithm for nonlinear optimization problems with binary variables. Computational Optimization and Applications. 2010; **47**(2):257-288

[9] Waltz RA, Morales JL, Nocedal J, et al. An interior algorithm for nonlinear optimization that combines line search and trust region steps. Mathematical Programming. 2006;**107**(3):391-408

[10] Nazemi A. A dynamic system model for solving convex nonlinear optimization problems. Communications in Nonlinear Science and Numerical Simulation. 2012;**17**(4):1696-1705

[11] Yang L, Shami A. On hyperparameter optimization of machine learning algorithms: Theory and practice. Neurocomputing. 2020; **415**:295-316

[12] Bao Y, Hu Z, Xiong T. A PSO and pattern search based memetic algorithm for SVMs parameters optimization. Neurocomputing. 2013;**117**:98-106

[13] Wang Z, So O, Lee K, et al. Adaptive Risk Sensitive Model Predictive Control with Stochastic Search. Vancouver: ML Research Press; 2021. pp. 510-522

[14] Andersson JA, Gillis J, Horn G, et al. CasADi: A software framework for nonlinear optimization and optimal control. Mathematical Programming Computation. 2019;**11**(1):1-36

[15] Mahajan A, Leyffer S, Linderoth J, et al. Minotaur: A mixed-integer nonlinear optimization toolkit. Mathematical Programming Computation. 2021;**13**(2):301-338

[16] Diehl M, Bock HG, Kostina E. An approximation technique for robust nonlinear optimization. Mathematical Programming. 2006;**107**(1):213-230

[17] Jahn J. Introduction to the Theory of Nonlinear Optimization. London: Springer Nature; 2020

[18] Hochbaum DS. Complexity and algorithms for nonlinear optimization problems. Annals of Operations Research. 2007;**153**(1):257-296

[19] Rusch T, Mair P, Hornik K. Cluster optimized proximity scaling. Journal of Computational and Graphical Statistics. 2021;**30**(4):1156-1167

[20] M'hamdi B, Teguar M, Mekhaldi A. Optimal design of corona ring on HV composite insulator using PSO approach with dynamic population size. IEEE Transactions on Dielectrics and Electrical Insulation. 2016;**23**(2): 1048-1057

[21] Sardashti A, Daniali H, Varedi S. Optimal free-defect synthesis of four-bar linkage with joint clearance using PSO algorithm. Meccanica. 2013;**48**(7): 1681-1693

[22] Xu F, Chen H, Gong X, et al. Fast nonlinear model predictive control on FPGA using particle swarm optimization. IEEE Transactions on Industrial Electronics. 2015;**63**(1):310-321

[23] Zhu H, Wang Y, Wang K, et al. Particle swarm optimization (PSO) for the constrained portfolio optimization problem. Expert Systems with Applications. 2011;**38**(8):10161-10169

[24] Zhou J, Fang R, Li Y, et al. Parameter optimization of nonlinear grey Bernoulli model using particle swarm optimization. Applied Mathematics and Computation. 2009;**207**(2): 292-299

[25] Hota P, Barisal A, Chakrabarti R. An improved PSO technique for short-term optimal hydrothermal scheduling. Electric Power Systems Research. 2009; **79**(7):1047-1053

[26] Krohling RA, dos Santos Coelho L. PSO-E: Particle Swarm with Exponential Distribution. Vancouver: ML Research Press; 2006. pp. 1428-1433

[27] Hosseinpourfard R, Javidi MM. Chaotic pso using the lorenz system: An efficient approach for optimizing nonlinear problems. Çankaya University Journal of Science and Engineering. 2015;**12**(1):40-59

[28] Song Y, Chen Z, Yuan Z. New chaotic PSO-based neural network predictive control for nonlinear process. IEEE Transactions on Neural Networks. 2007;**18**(2):595-601

[29] Panda S, Mohanty B, Hota PK. Hybrid BFOA–PSO algorithm for automatic generation control of linear and nonlinear interconnected power systems. Applied Soft Computing. 2013; **13**(12):4718-4730

[30] Luo S, Liang W, Zhao G. Hybrid PSO-WDBA method for the site selection of tailings pond. Computers & Industrial Engineering. 2020;**143**:106429

[31] Armand P, Omheni R. A mixed logarithmic barrier-augmented Lagrangian method for nonlinear optimization. Journal of Optimization Theory and Applications. 2017;**173**(2): 523-547

[32] Guo D, Yan L, Nie Z. Design, analysis, and representation of novel five-step DTZD algorithm for time-varying nonlinear optimization. IEEE Transactions on Neural Networks and Learning Systems. 2017;**29**(9): 4248-4260

[33] Milz J, Ulbrich M. An approximation scheme for distributionally robust nonlinear optimization. SIAM Journal on Optimization. 2020;**30**(3):1996-2025

[34] Wang S, Chiang H-D. Constrained multiobjective nonlinear optimization: A user preference enabling method. IEEE Transactions on Cybernetics. 2018; **49**(7):2779-2791

[35] Ding J, Wang Q, Zhang Q, et al. A hybrid particle swarm optimization-cuckoo search algorithm and its engineering applications. Mathematical Problems in Engineering. 2019;**2019**:1-12

[36] Zhang J, Xia P. An improved PSO algorithm for parameter identification of nonlinear dynamic hysteretic models. Journal of Sound and Vibration. 2017; **389**:153-167

[37] Gao G, Liu F, San H, et al. Hybrid optimal kinematic parameter identification for an industrial robot based on BPNN-PSO. Complexity. 2018; **2018**:1-11

[38] El-Shorbagy MA, Elhoseny M, Hassanien AE, et al. A novel PSO algorithm for dynamic wireless sensor network multiobjective optimization problem. Transactions on Emerging Telecommunications Technologies. 2019;**30**(11):e3523

[39] Godio A, Santilano A. On the optimization of electromagnetic geophysical data: Application of the PSO algorithm. Journal of Applied Geophysics. 2018;**148**:163-174

[40] Song W, Cattani C, Chi C-H. Multifractional Brownian motion and quantum-behaved particle swarm optimization for short term power load forecasting: An integrated approach. Energy. 2020;**194**:1-8

[41] Xiaohong Q, Xiaohui Q. An evolutionary particle swarm optimizer based on fractal brownian motion. Journal of Computational Intelligence and Electronic Systems. 2012;**1**:138-143

[42] Du J, Liu Y, Yu Y, et al. A prediction of precipitation data based on support vector machine and particle swarm optimization (PSO-SVM) algorithms. Algorithms. 2017;**10**(2):57

[43] Zapotecas-Martínez S, Coello CAC, Aguirre HE, et al. A review of features and limitations of existing scalable multiobjective test suites. IEEE Transactions on Evolutionary Computation. 2018;**23**(1):130-142

[44] Zhang W, Li G, Zhang W, et al. A cluster based PSO with leader updating mechanism and ring-topology for multimodal multi-objective optimization. Swarm and Evolutionary Computation. 2019;**50**:100569

[45] Nadeau PC. Multiobjective Nelder-Mead Algorithm Using a Mesh-Map of Weighted Sums. Vancouver: University of British Columbia; 2020

[46] Lin T, Chen T, Liu J, Tu XM. Extending the Mann-Whitney-Wilcoxon rank sum test to survey data for comparing mean ranks. Statistics in Medicine. 2021;**40**:1705-1717

[47] de Barros RSM, Hidalgo JIG, de Lima Cabral DR. Wilcoxon rank sum test drift detector. Neurocomputing. 2018;**275**: 1954-1963

Chapter 3

Bi-Level Non-deterministic Optimization Methods Applied in Process Synthesis

Leandro V. Pavão, Caliane B.B. Costa and Mauro A.S.S. Ravagnani

Abstract

In this topic, bi-level optimization methods based on non-deterministic approaches are presented with emphasis on solving process synthesis problems, in large-scale problems, where determinist approaches can lead to non-optimal solutions, with huge computational effort. When integer and real variables are present, bi-level approaches can present good solutions in reasonable computational times. Normally, binary variables are used when decisions must be taken, such as whether to use a piece of equipment, whether a process stream must be split, etc. Real variables are the other design variables. Some case studies are used to show the applicability of these methods. The chosen examples are related to process synthesis and heat integration processes, with distinct applications.

Keywords: optimization, bi-level optimization methods, non-deterministic approaches, process synthesis, heat integration

1. Introduction

In industrial processes, optimization techniques have been used to improve the process performance and reduce costs and environmental impacts. Different optimization approaches have been used, like deterministic, stochastic, and hybrid ones, combining both strategies.

In recent papers [1–3], the authors provided complete reviews of the progress in optimization-based process synthesis problems. Important advances in optimization and tools in process synthesis were presented, like machine learning, quantum computing as well as assisted approaches. Challenges in the theory and practice in process synthesis, and the revision of important concepts in conceptual design and process design, based on superstructures, and modeling environments, for the major subsystems and flowsheets are discussed. Novel and current features for sustainable energy production processes, such as biomass transformation, shale production, and intermittent production of energy, were also discussed.

Specific techniques to solve the optimization problems were proposed in the literature. Depending on the size of the optimization problem, decomposition techniques can be proposed to achieve the problem solution. Bi-level optimization approaches are one of the most used strategies to circumvent the original difficulties in solving large-scale problems.

A bi-level problem can be defined as an optimization problem in which a subset of variables is constrained to be optimal for another given problem parameterized by the remaining variables. The outer problem is commonly referred to as the master problem, and the inner one is the slave problem.

Specifically, in the field of process synthesis, even when small-scale problems are formulated as optimization problems, finding good solutions can be quite a difficult task. Problems involving heat and mass integration, heat and work integration, and distillation sequencing with thermal coupling are regular problems with high degrees of difficulty and are a promising field for the application of problem decomposition strategies, such as bi-level optimization strategies. In the same way, superstructure-based optimization techniques are the most used when decomposition approaches are proposed for the problem solution.

Heat exchanger networks are an active research field and for more than 50 years, a large number of papers have been dedicated to present alternative techniques to find good and optimal solutions for the problem. Due to the increasing global competition and the necessity to decrease the harmful effects on the environment, it will be a continuous expansion field in order to improve the heat exchanger networks and their synthesizing methods. The synthesis of heat exchanger networks can involve several optimization tasks, such as the definition of stream matches, considering stream-splitting, and the distribution of heat loads. The optimization of the network depends also on well-designed heat exchangers, which enable heat recovery in process industries in an energy-, environment-, and cost-efficient manner.

Because of these challenges, the synthesis and optimization of heat exchanger networks is still an open problem due to the complexity of the search space that is comprised of possible solutions. Moreover, most of the proposed methods present simplifying assumptions that influence the topological features of the candidate solutions considered, limiting the boundaries of the search space.

In the literature, authors treated bi-level programming problems with distinct approaches. In Ref. [4], novel algorithms for the global solution of bi-level mixed-integer linear programming and mixed-integer quadratic programming problems with both integer and continuous variables were considered. In Ref. [5], global optimization approaches were proposed to solve nonlinear bi-level optimization problems. Authors used the Branch and Bound αBB global optimization algorithm jointly with relaxation techniques applied to the search space in order to achieve convex underestimations, in a finite number of iterations to ensure the global optimality of the problem.

Laukkanen and Fogelholm [6] presented a bi-level optimization method for the simultaneous synthesis of heat exchanger networks using a grouping of streams, aggregate streams, and bi-level optimization to decompose the set of binary variables (the variables that define the existence of heat exchange matches) into two separate problems. They found better results when compared to the solutions presented in the literature, using the same superstructure, with reduced calculation effort for large-scale problems. In another work of the same research group [7] the authors combined an interactive multi-objective optimization method with the simultaneous bi-level HENS method, using the proposition presented in Ref. [6].

In Ref. [8], the authors used the advantages of the sequential and simultaneous approaches for the optimal synthesis of heat exchanger networks. The proposed approach divides the problem into two sequential parts. The first part uses the transportation model and the concept of temperature intervals. The second step uses a superstructure containing all the possible alternatives in which the matches predicted by the first stage can exchange heat to synthesize the final heat exchanger network.

Other papers presented the synthesis of heat exchanger networks combined with other heat exchanger features, like pressure drops [9] and the detailed equipment design [10], in a bi-level approach. On the outer level, a network topology is optimized and on the inner level, the other calculations were provided. Other interesting features that were combined to heat exchanger network synthesis were work integration [11], where not only heating and cooling are necessary, but also compression and expansion are required by streams, and heat integration of distillation sequences [12], where sequences need to be devised simultaneously to their heat integration system.

The multiperiod synthesis and optimization of integrated heat exchanger networks and Rankine cycles is also a challenging optimization problem. In Ref. [13], the authors provided a general and systematic synthesis methodology to optimize simultaneously the utility systems, Rankine cycles, and the heat exchanger networks synthesis. The problem has the formulation of a nonconvex mixed-integer nonlinear programming (MINLP) problem and was solved with a bi-level decomposition method.

Meta-heuristic approaches have also been used to achieve good solutions in bi-level optimization problems for the heat exchanger network synthesis. Several meta-heuristic approaches have been proposed in the literature. In Ref. [14], the authors presented a bi-level algorithm for the optimal heat exchanger network synthesis, optimizing separately the integer and the continuous variables. The outer level is a problem-oriented evolution method, which generates new candidate structures. The inner level is a memetic particle swarm optimization (PSO) algorithm, an improved particle swarm optimization combined with a local search component.

In Ref. [15], a bi-level no-split heat exchanger network synthesis hybrid method was presented. On the outer level, simulated annealing (SA) was used to optimize the network topology. The continuous variables were handled using a PSO algorithm. A parallel processing technique was also presented to improve local search performance.

A bi-level optimization approach was proposed by Aguitoni et al. [16] for the synthesis of heat exchanger networks. A hybrid meta-heuristic method combining SA and differential evolution (DE) was introduced. SA was used to optimize the network topologies and DE was applied to the continuous variables.

In the next sections, we present some methods based on non-deterministic approaches for process synthesis problems, considering the synthesis of heat exchanger networks, heat and work exchange networks, and heat-integrated distillation sequences. Large-scale problems were considered, where determinist approaches can lead to non-optimal solutions, with huge computational effort.

2. Method

Bi-level optimization using meta-heuristics consists of merging capabilities of two types of stochastic optimization methods. What is, in general, called "outer level" method encompasses a meta-heuristic for combinatory optimization such as Simulated Annealing (SA) [17], Tabu Search (TS) [18], or Genetic Algorithms (GA). That method is applied to disturb discrete variables present in the optimization model. In the context of process systems engineering, these variables in general may represent

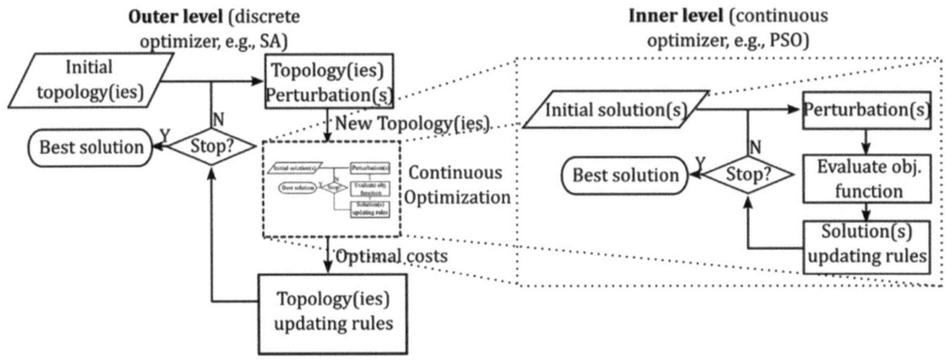

Figure 1.
Schematic representation of bi-level approach (adapted from [15]).

the presence or absence of unit operations, bypasses, components, etc., in a process. On the other hand, the "inner level" represents continuous variables that are associated with the discrete variables that are enabled in the "outer level." For instance, if a heat exchanger exists, one can find a heat load value so that an optimal compromise between energy recovery and capital cost of that unit is found. Methods used to optimize inner-level variables are, for instance, particle swarm optimization (PSO) [19] and differential evolution (DE) [20]. **Figure 1** presents a general schematic representation of a bi-level meta-heuristic optimization approach.

In terms of mathematical programming, bi-level approaches are suitable for the optimization of mixed-integer nonlinear programming problems. The minimization case can be generally formulated as follows:

$$\begin{array}{ll} \min & \{f(x,b)\} \\ s.a. & h(x,b)=0 \\ & g(x,b)\leq 0 \\ & x\in X\subseteq \mathbb{R}^d, b\in B\subseteq \mathbb{Z}^q \end{array} \quad (1)$$

where f is an objective function being minimized, h are equality constraints, and g the inequality ones. In a bi-level approach, the outer level would comprise altering the binary variables (b, which has q dimensions), while continuous ones (x, which has d dimensions) would be changed by the inner-level method.

3. Optimization methods

Some of the most prominent methods used in bi-level approaches are simulated annealing (SA) and particle swarm optimization (PSO). Some background on these will be provided here.

Simulated annealing was proposed by Kirkpatrick et al. [17]. It uses an analogy to the annealing process of solids where one should aim for low energy states with properly slow temperature reduction. In SA, the optimization method starts with a single solution whose neighborhood is explored with small moves at each iteration. When a better solution is found, the method automatically accepts it as the current solution. After a given number of iterations, a temperature parameter is reduced.

Temperature, in this context, is associated with the probability of a "worse" solution being accepted as the current one according to the following equation:

$$PoA(\Delta fobj, T) = e^{-\frac{\Delta fobj}{T}} \tag{2}$$

where $\Delta f\, obj$ represents the difference between the objective function values for the new solution and the current solution. This probability is calculated only when the new proposed solution has a worse objective function value than the current one.

This strategy aids the method in escaping from local minima. Larger temperatures yield greater probabilities of acceptance, meaning broader exploration. The method is terminated after a minimum temperature value is met. Note that SA can be readily adapted to continuous spaces as well by replacing neighborhood moves with random exploration of continuous variables.

Particle Swarm Optimization was proposed by Kennedy and Eberhart [19] and is a widely used continuous optimization approach. The method draws inspiration from the collective behavior of animal groups, such as swarms or flocks, as they search for optimal resources, such as food-rich locations. PSO takes advantage of both collective and cognitive learning in defining "particle" behavior iteratively.

Mathematically, PSO performs two simple calculations at each iteration, which are as follows:

$$v_{k+1}^{(p)} = \omega v_k^{(p)} + c_1 r_1 \left(p_k^{(p)} - x_k^{(p)}\right) + c_2 r_2 \left(p_k^{(global)} - x_k^{(p)}\right) \tag{3}$$

$$x_{k+1}^{(i)} = x_k^{(i)} + v_{k+1}^{(i)} \tag{4}$$

where x denotes the current position of particle p at a given k or $k + 1$ iteration, v represents the particle's velocity, $p^{(p)}$ denotes the best position found by a given particle, $p^{(global)}$ denotes the best position found by the swarm, c_1 and c_2 represent cognitive and social parameters, r_1 and r_2 are random variables, and ω is the inertia parameter.

4. Process systems engineering and bi-level approaches

In the context of process systems, several problems can be tackled *via* bi-level methods. This section presents some more in-depth rationale behind the use of these methods in process systems engineering problems. Namely, models for the synthesis of (i) heat exchanger networks, (ii) heat and work exchange networks, and (iii) heat-integrated distillation sequences will be presented.

4.1 Heat exchanger networks (HEN)

One of the most employed methods for modeling the HEN synthesis problem is the stage-wise superstructure (SWS) proposed by Yee and Grossmann [21]. The idea of the superstructure is to, in a given stage (multiple stages can be present), represent all possible matches between streams. In those matches, heat is recovered, typically in counter-current heat exchangers. Each match is present in an exclusive stream split branch. In the original model derived from the SWS by Yee and Grossmann, the authors assumed isothermal mixing, which simplifies the model and fixes values for stream split fractions as functions of heat loads of matches present at respective branches. The model can be broadened to handle stream split fractions as decision variables as well. The superstructure is presented in **Figure 2**.

A MINLP model can be derived from the SWS considering non-isothermal mixing. The minimization model equation is presented as follows and can provide a general notion of the notation used and how to relate it to the stage-wise superstructure. It is based on the work of Pavão et al. [22], and the complete version with in-depth description of each equation can be found there, including energy balances and heat exchange area calculation equations:

$$\begin{aligned} min \quad & \sum_i C_{cu} \bullet Qcu_i + \sum_j C_{hu} \bullet Qhu_j + \\ & \sum_i \sum_j \sum_k z_{i,j,k} \bullet \left(B + C \bullet A_{i,j,k}{}^{\beta}\right) + \\ & \sum_i zcu_i \bullet \left(B + C \bullet Acu_i{}^{\beta}\right) + \qquad i \in N_H, j \in N_C, k \in N_S \qquad (5) \\ & \sum_j zhu_j \bullet \left(B + C \bullet Ahu_j{}^{\beta}\right) \\ s.t. \quad & \text{HEN constraints} \end{aligned}$$

where the first two terms represent operating costs and the third to fifth represent capital costs. *Ccu* and *Chu* represent operating costs for cold and hot utilities in general given in monetary values per energy units (e.g., \$/kWy and \$./GJ), *Qcu* and *Qhu* are coolers and heaters heat loads. Note that these terms have *i* and *j* indexes, respectively, implying that coolers are only applied to the ends of hot streams and heaters to the ends of cold streams. Evidently, more complex models can be created to consider the allocation of heaters or coolers at intermediate positions [23]. The third term comprises capital costs for heat recovery units. It has a *z* binary variable that denotes the presence/absence of heat exchangers and whose indexes (i, j, k) denote that a match is between hot stream *i* and cold stream *j* in stage *k*. These binary variables are associated with a function of the heat exchanger area (A) and that yields capital costs of the heat exchanger. The capital costs formula used here has a fixed costs constant (B), a variable cost constant (C), and a capital costs exponent (β). The fourth and fifth terms are for the capital costs of coolers and heaters. They also have terms that represent the presence/absence of those units (*zcu* and *zhu*) and multiply cost functions similar to the aforementioned.

In the context of bi-level optimization, the variables representing the presence/absence of heat recovery units (z) can be activated or deactivated using a

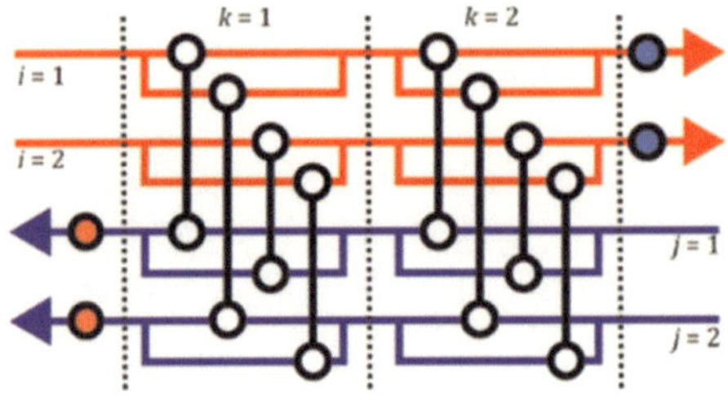

Figure 2.
Stage-wise superstructure [21] for two hot and two cold streams.

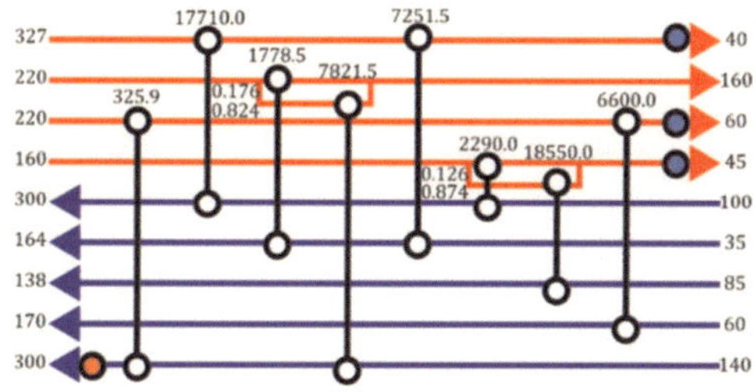

Figure 3.
Final configuration for 9-streams aromatics plant problem (amended from [22]).

combinatorial optimization meta-heuristic method such as Simulated Annealing. In that case, each SA move consists of adding or removing a heat exchanger from the network structure, giving rise to a new HEN topology. That new topology has associated heat loads and, if non-isothermal mixing is considered, stream split fractions. These decision variables can be optimized with an inner-level continuous optimization approach such as PSO. As in Pavão et al. [22], a hybrid continuous optimization method can be employed as well: a continuous SA adaptation performs initial scouting throughout the search area, and its final solution is incorporated into the initial swarm in PSO, providing the method with a promising solution for refinement.

An interesting result of this method [22] application can be presented here. It regards the HEN synthesis for the classic 9-streams aromatics plant. The TAC achieved by this methodology was 2,919,675 $/y, which is a competitive solution in comparison to the literature, especially if only works that employ models derived from similar stage-wise superstructure are observed. **Figure 3** presents the mentioned solution.

4.2 Work and heat exchange networks (WHEN)

Work and heat exchange network synthesis is a relatively new field, in contrast to the well-studied HEN synthesis problem. It can be considered an extension of the former field by including not only streams suitable for heat recovery but also streams needing compression or expansion (*via* compressors, turbines, or valves). In that case, streams can be also suitable for some form of work recovery, for instance, through single-shaft-turbine-compressor couplings, which may or may not require auxiliary motors or generators. Many approaches, in fact, take as basis several concepts used in HEN synthesis and include the possibility of compression or expansion.

The procedure presented here for WHEN synthesis takes advantage of the stage-wise superstructure for its heat recovery stage. However, a work recovery stage is also present. It is worth noting that compressors/turbines/valves not only change stream pressures but also the inlet temperatures of streams that undergo heat recovery, which makes the problem even more complex. **Figure 4** illustrates the expanded superstructure that is used here.

Note that in the WHEN superstructure, there is a general w index. It is a stream index used both for hot and cold streams. It is interpreted within the algorithm code to an i, j form as in the HEN model for heat recovery calculations, while compression/expansion units keep the general form. These units are blocks, as shown in **Figure 4b**, that may be plugged into the slots at each stream work recovery section. The model described here is devised from that superstructure and is based on the set of equations

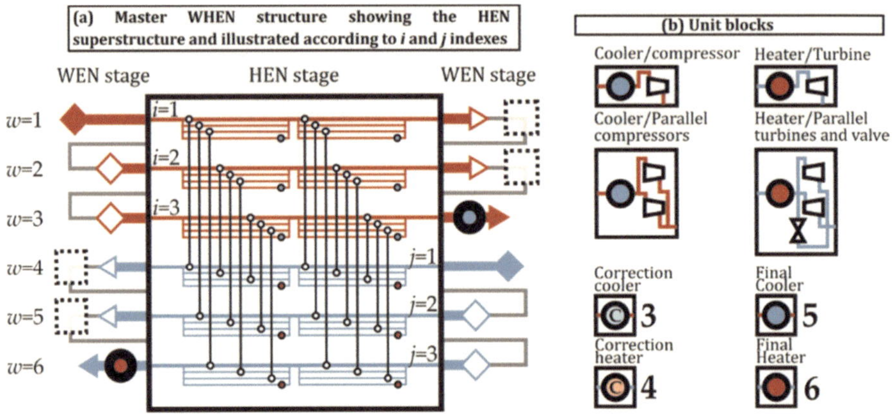

Figure 4.
(a) Expanded stage-wise superstructure for work and heat recovery and (b) blocks that are used in the WEN stage (adapted from [24]).

devised in the work of Pavão et al. [24]. Other equations and in-depth explanations can be found in the referred work.

Operating costs are calculated as follows:

$$
\begin{aligned}
OC = & \left(\sum_{m \in NHU} \sum_{j \in NC} \sum_{k \in NS} zhuinter_{i,m,k} Chu_m \cdot Qhuinter_{m,j,k} + \sum_{m \in NHU} \sum_{oj \in NOC} zhu_m Chu_m Qhu_{oj} \right) + \\
& \left(\sum_{i \in NH} \sum_{n \in NCU} \sum_{k \in NS} zcuinter_{i,n,k} Ccu_n \cdot Qcuinter_{i,n,k} + \sum_{n \in NCU} \sum_{oi \in NOH} zcu_n Ccu_n Qcu_{oi} \right) + \\
& Cel \cdot \left(TotalCompWork + \sum_{s \in NSh} SSTCWorkLack_s \right) - \\
& Rel \cdot \left(TotalTurbWork + \sum_{s \in NSh} SSTCWorkSurplus_s \right)
\end{aligned}
\tag{6}
$$

where *Chu* and *Ccu* are for hot and cold utility costs. Note that *m* and *n* are indexes for different utility type costs. The utility type is decided by the discrete optimizer by changing the values of *zhuinter* and *zcuinter*. *Qhuinter* and *Qcuinter* are heat loads of heaters and coolers placed at intermediate positions in the superstructure. *Qhu* and *Qcu* are total hot and cold utilities that are yet needed in the stream end after all passes through the heat recovery area. Note that their indexes (*oi* and *oj*) are for original hot and cold streams and not a specific pass. The variables *zhu* and *zcu* determine what type of utility is used in those streams. Moreover, operating costs related to compression/expansion are present. *Cel* and *Rel* are electricity costs and revenue prices. *TotalCompWork* and *TotalTurbWork* are the sum of all work rate required or produced in compressors or turbines. *SSTCWorkLack* is the work required in an *s* SSTC coupling (that is, if turbines in a shaft cannot provide the work required by compressors in that shaft, an auxiliary motor is used to provide such power). The analogous is valid for *SSTCWorkSurplus*, where a generator is used when the work needed by compressors is lower than that produced by the turbines. A generator produces electricity, which is sold to the grid.

The calculation of capital costs is performed as follows:

$$
\begin{aligned}
CC = & \sum_{i \in NH} \sum_{j \in NC} \sum_{k \in NS} AreaCost(A_{i,j,k}) + \sum_{i \in NH} \sum_{n \in NCU} \sum_{k \in NS} AreaCost(Acuinter_{i,n,k}) \\
& + \sum_{m \in NHU} \sum_{j \in NC} \sum_{k \in NS} AreaCost(Ahuinter_{m,j,k}) + \sum_{w \in NG} AreaCost(Acu_w) \\
& + \sum_{w \in NG} AreaCost(Ahu_w) + \sum_{w \in NG} AreaCost(Apre_w) + \sum_{w \in NG} AreaCost(Acor_w) \\
& + \sum_{w \in NG} \sum_{f \in NF} CompCost_{w,f} + \sum_{w \in NG} \sum_{f \in NF} ExpCost_{w,f} + \sum_{s \in SS} MotCost_s + \sum_{s \in SS} GenCost_s
\end{aligned}
\tag{7}
$$

where f is an index for different parallel compressors, turbines, or valves in the blocks for the work recovery section. The optimization model can be formulated as the minimization of the sum of annualized capital and operating costs.

The optimization method for WHEN in this case uses a similar approach to that described for HEN synthesis: SA on the outer level and the CSA/PSO hybrid on the inner level. SA adds or removes heat exchangers, compression, or expansion units, while CSA/PSO optimizes decision variables associated with them (heat/workloads). Moreover, the derived model [24] is able to find solutions that outperform deterministic approaches, leading to TAC of 9,752,857 \$/y when the number of couplings is not limited, neither are compressor/turbine discharge pressures. When practical constraints are included (maximum of three coupled units per shaft and discharge temperatures bounded to a practical level), TAC slightly increases to 9,873,282 \$/y but provides more technically realistic results. Works from the literature present TAC of 10.5×10^6 and 10.2×10^6 \$/y [25]. **Figure 5** presents solutions found for both aforementioned scenarios.

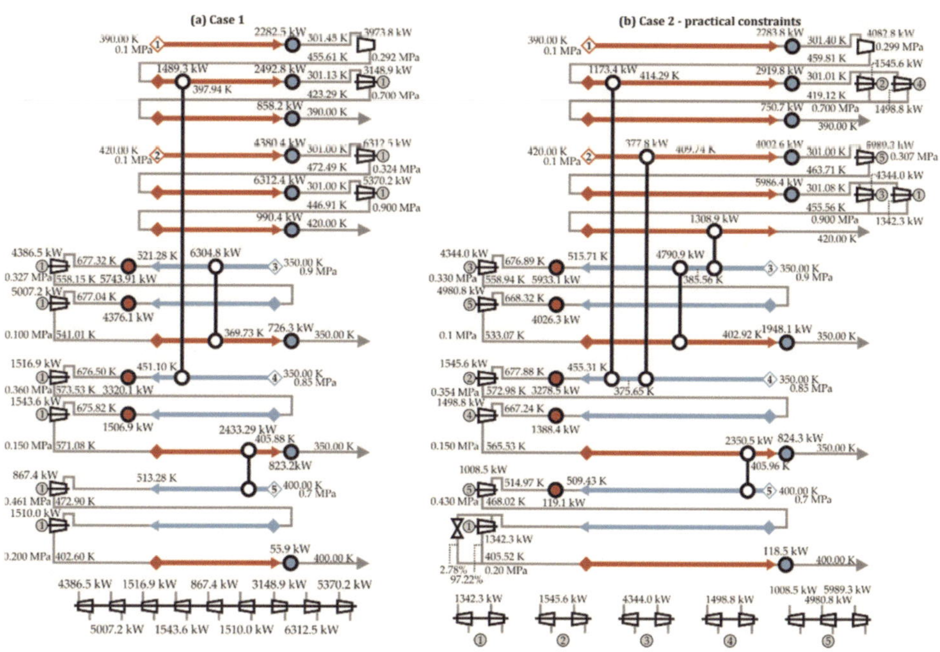

Figure 5.
Solutions for the (a) base case and for the (b) scenario with practical constraints (amended from Ref. [24]).

4.3 Heat-integrated distillation sequences (HIDiS)

The synthesis of multicomponent distillation sequences involves choosing in what order components are removed as distillate or bottom products and at what conditions such as pressures and reflux ratios those separations occur. Moreover, the number of equilibrium stages to perform each separation must be defined as well as the optimum feed stage at each column. Typically, a total annual cost function can be minimized considering capital costs for columns, reboilers, and condensers, as well as operating costs associated with heat loads in reboilers and condensers. Heat-integrated distillation sequences are an extension to that problem, and one should consider the possibility of recovering heat from hot distillate streams to cold bottom streams. Component separation order, reflux ratios, and especially column pressures can be manipulated in such a manner that heat recovery is favored by changing top and bottom temperatures.

One way to perform such a task is to perform bi-level optimization in two stages: (i) the distillation sequence synthesis, without considering heat integration, or considering it implicitly through a heuristic manner (e.g., Pinch Analysis supertargeting) and (ii) conducting explicit HEN synthesis to the streams available in the previously devised distillation sequence.

Figure 6 presents a superstructure for HIDiS synthesis that can give rise to an optimization model. In a bi-level optimization approach, columns can be added/ removed to the available "slots" and key components can be changed by a discrete optimizer, while pressures and reflux ratios can be defined with a continuous optimizer.

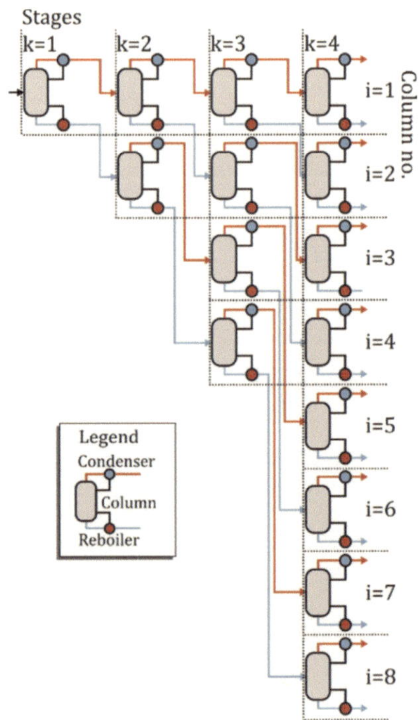

Figure 6.
HIDiS superstructure (amended from Ref. [26]).

A mathematical model can be derived on a staged basis for this problem. Columns can be calculated using shortcut methods, such as the Fenske-Underwood-Gilliland equations. The following equations present operating and capital cost calculations for the distillation sequence (without heat integration) and are based on the derivation proposed by Pavão et al. [26]:

$$OCDiS = \sum_{k \in ST} \sum_{i \in COLS} \left(\sum_{p \in HU} (zhu_{p,i,k} Chu_p \cdot Qrebcol_{i,k}) + \sum_{q \in CU} (zcu_{q,i,k} Ccu_q \cdot Qcondcol_{i,k}) \right) \tag{8}$$

$$CCDiS = \sum_{k \in ST} \sum_{i \in COLS} (CCcol_{i,k} + CCcond_{i,k} + CCreb_{i,k}) \tag{9}$$

where zhu and zcu are binary variables that indicate what is the utility type used in reboilers and condensers (p and q indexes) at column i at stage k. $Qrebcol$ and $Qcondcol$ are heat loads for reboilers and condensers. $CCcol$, $CCcond$, and $CCreb$ are capital costs for columns, condensers, and reboilers and are calculated using specific capital costs formulations. A simple formulation is described by Pavão et al. [26].

When heat integration is performed implicitly, the HIDiS capital costs consist of column-related costs only, while condenser/reboiler costs are replaced by the predicted area-related costs from the Pinch Analysis. In the HIDiS operating costs, condenser/reboiler heat load-related costs are replaced by operating costs from Pinch Analysis. When explicit heat integration is performed, OC and CC from the synthesized heat exchanger network are inserted instead of those predicted by Pinch Analysis. The optimization model consists thus of minimizing TAC ($OC + CC$).

In the example presented here based on Ref. [26], the bi-level optimization is performed in two stages: (i) HIDiS with implicit heat integration and (ii) explicit heat integration. At first, an outer level SA proposes new DiS structures by adding/removing columns or slot bypasses, changing key components for separation, and also performing "column stacking," which uses the FUG method to simulate columns with multiple feeds and side draws and also eliminates some reboilers and condensers. Column stacking can be performed for columns in a single stage. Given the structure proposed by SA, a hybrid CSA/PSO method modifies (i) column pressures, (ii) reflux ratios, (iii) heat recovery approach temperature ($HRAT$, a parameter for Pinch Analysis that denotes a "degree" of heat integration and whose compromise with heat exchange area must be evaluated), and (iv) fractions of use of hot and cold utility types (over total utility requirements). In the explicit heat integration step, the method extracts process streams from condensers and reboilers from the previous

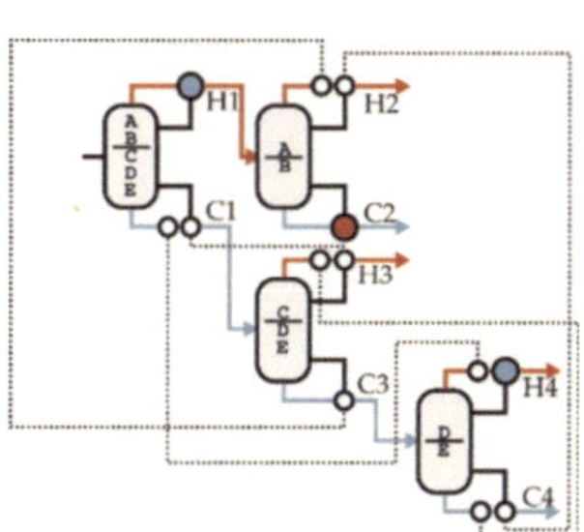

Figure 7.
Result found for alcohol separation problem (amended from Ref. [26]).

step. It then synthesizes a HEN using a bi-level methodology similar to that proposed in Section 3.1.

Figure 7 demonstrates results for a separation problem where a 500.4 kmol/h mixture of five alcohols (0.25 ethanol, 0.15 i-propanol, 0.35 n-propanol, 0.10 i-butanol, and 0.15 n-butanol) must be separated to nearly pure streams.

That solution has a TAC of 332,698 $/y, and it outperforms previous literature (379,231 $/y [27]).

5. Conclusions

Bi-level optimization has proven to be an effective strategy for addressing complex process synthesis problems, particularly in large-scale systems where traditional methods may not be efficient. By combining meta-heuristic approaches for discrete decisions such as the presence/absence of pieces of equipment with continuous optimization techniques for optimization of process variables such as heat loads or operating pressures, bi-level framework performance is promising in computational challenges inherent in mixed-integer nonlinear programming. Applications in computationally demanding problems from process systems engineering—heat exchanger network synthesis, work and heat exchange networks, and heat-integrated distillation sequences—illustrate the versatility and efficiency of these methods in achieving competitive solutions while aiming for optimal trade-offs between energy recovery, cost efficiency, and handling complex operational constraints. Despite the success achieved so far with bi-level approaches, challenges remain, including simplifying assumptions that limit the exploration of optimal solutions and the computational intensity of detailed model optimization. Future efforts should aim to expand the scope of bi-level methods toward more complex processes, possibly integrating process simulators and surrogate models based on artificial intelligence approaches to the models for efficiency. With its ability to manage discrete and continuous variables effectively, bi-level optimization has become an important tool for advancing sustainable and efficient process engineering solutions.

Acknowledgements

The authors acknowledge financial support from the Conselho Nacional de Desenvolvimento Científico e Tecnológico—Brasil (CNPq), processes 307958/2021-3, 309026/2022-9, and 406544/2023-9, and the Coordenação de Aperfeiçoamento de Pessoal de Nível Superior—Brasil (CAPES)—Finance Code 001.

Conflict of interest

The authors declare no conflict of interest.

Author details

Leandro V. Pavão, Caliane B.B. Costa and Mauro A.S.S. Ravagnani*
Chemical Engineering Graduate Program, State University of Maringá, Maringá, PR, Brazil

*Address all correspondence to: massravagnani@uem.br

References

[1] Pistikopoulos EN, Tian Y. Advanced modeling and optimization strategies for process synthesis. Annual Review of Chemical and Biomolecular Engineering. 2024;**15**:81-103. DOI: 10.1146/annurev-chembioeng-100522-112139

[2] Chen Q, Grossmann IE. Recent developments and challenges in optimization-based process synthesis. Annual Review of Chemical and Biomolecular Engineering. 2017;7(8): 249-283. DOI: 10.1146/annurev-chembioeng-080615-033546

[3] Montastruc L, Belletante S, Pagot A, Negny S, Raynal L. From conceptual design to process design optimization: A review on flowsheet synthesis. Oil & Gas Science and Technology - Rev. IFP Energies Nouvelles. 2019;**74**:80. DOI: 10.2516/ogst/2019048

[4] Avraamidou S, Pistikopoulos EN. A multi-parametric optimization approach for bilevel mixed-integer linear and quadratic programming problems. Computers & Chemical Engineering. 2019; 125;(9):98-113. DOI: 10.1016/j.compchemeng.2019.01.021

[5] Gümüş ZH, Floudas CA. Global optimization of nonlinear bilevel programming problems. Journal of Global Optimization. 2001;**2001**(20): 1-31. DOI: 10.1023/A:1011268113791

[6] Laukkanen T, Fogelholm CJ. A bilevel optimization method for simultaneous synthesis of medium-scale heat exchanger networks based on grouping of process streams. Computers & Chemical Engineering. 2011;**35**(11): 2389-2400. DOI: 10.1016/j.compchemeng.2010.11.009

[7] Laukkanen T, Tveit TM, Ojalehto V, Miettinen K, Fogelholm CJ. Bilevel heat exchanger network synthesis with an interactive multi-objective optimization method. Applied Thermal Engineering. 2012;**48**(15):301-316. DOI: 10.1016/j.applthermaleng.2012.04.058

[8] Caballero JA, Pavão LV, Costa CBB, Ravagnani MASS. A novel sequential approach for the design of heat exchanger networks. Frontiers in Chemical Engineering. 2021;**3**:733186. DOI: 10.3389/fceng.2021.733186

[9] Serna-González M, Ponce-Ortega JM, Jiménez-Gutierrez A. Two-level optimization algorithm for heat exchanger networks including pressure drop considerations. Industrial & Engineering Chemistry Research. 2004;**43**(21): 6766-6773. DOI: 10.1021/ie0497700

[10] Ravagnani MASS, Caballero JA. Optimal heat exchanger network synthesis with the detailed heat transfer equipment design. Computers & Chemical Engineering. 2007;**31**(11): 1432-1448. DOI: 10.1016/j.compchemeng.2006.12.005

[11] Onishi VC, Ravagnani MASS, Caballero JA. Simultaneous synthesis of work exchange networks with heat integration. Chemical Engineering Science. 2014;**112**:87-107. DOI: 10.1016/j.ces.2014.03.018

[12] Caballero JA, Grossmann IE. Structural considerations and modeling in the synthesis of heat-integrated–thermally coupled distillation sequences. Industrial & Engineering Chemistry Research. 2006; **45**(25):8454-8474. DOI: 10.1021/ie060030w

[13] Elsido C, Martelli E, Grossmann E I. Simultaneous multiperiod optimization of rankine cycles and heat exchanger networks. In: Proceedings of the 30th European Symposium on Computer

Aided Process Engineering (ESCAPE30), 24–27 May 2020, Milano, Italy

[14] Wang J, Cui G, Xiao Y, Luo X, Kabelac S. Bi-level heat exchanger network synthesis with evolution method for structure optimization and memetic particle swarm optimization for parameter optimization. Engineering Optimization. 2016;**49**(3):401-416. DOI: 10.1080/0305215X.2016.1191803

[15] Pavão L, Costa CBB, Ravagnani MASS. Heat exchanger network synthesis without stream splits using parallelized and simplified simulated annealing and particle swarm optimization. Chemical Engineering Science;**158**(2):96-107. DOI: 10.1016/j.ces.2016.09.030

[16] Aguitoni MC, Pavão LV, Ravagnani MASS. Heat exchanger network synthesis combining simulated annealing and differential evolution. Energy. 2019;**181**:654-664. DOI: 10.1016/j.energy.2019.05.211

[17] Kirkpatrick S, Gelatt CD, Vecchi MP. Optimization by simulated annealing. Science. 1983;**220**(4598):671-680. DOI: 10.1126/science.220.4598.671

[18] Glover F. Tabu search—Part I. INFORMS Journal on Computing. 1989;**1**(3):190-206. DOI: 10.1287/ijoc.1.3.190

[19] Kennedy J, Eberhart R. Particle swarm optimization. In: IEEE International Conference on Neural Networks. 1995. pp. 1942-1948. DOI: 10.1109/ICNN.1995.488968

[20] Storn R, Price K. Differential evolution–a simple and efficient heuristic for global optimization over continuous spaces. Journal of Global Optimization. 1997;**11**(4):341-359. DOI: 10.1023/A:1008202821328

[21] Yee TF, Grossmann IE. Simultaneous optimization models for heat integration —II. Heat exchanger network synthesis. Computers & Chemical Engineering. 1990;**14**(10):1165-1184. DOI: 10.1016/0098-1354(90)85010-8

[22] Pavão LV, Costa CBB, Ravagnani MASS, Jiménez L. Large-scale heat exchanger networks synthesis using simulated annealing and the novel rocket fireworks optimization. AICHE Journal. 2017;**63**(5):1582-1601. DOI: 10.1002/aic.15524

[23] Pavão LV, Costa CBB, Ravagnani MASS. An enhanced stage-wise superstructure for heat exchanger networks synthesis with new options for heaters and coolers placement. Industrial & Engineering Chemistry Research. 2018;**57**(7):2560-2573. DOI: 10.1021/acs.iecr.7b03336

[24] Pavão LV, Caballero JA, Ravagnani MASS, Costa CBB. An extended method for work and heat integration considering practical operating constraints. Energy Conversion & Management. 2020:206. DOI: 10.1016/j.enconman.2020.112469

[25] Huang K, Karimi IA. Work-heat exchanger network synthesis (WHENS). Energy. 2016;**113**:1006-1017. DOI: 10.1016/j.energy.2016.07.124

[26] Pavão LV, Caballero JA, Costa CBB, Ravagnani MASS. Synthesis of heat-integrated distillation sequences with nonsharp splits using a sequential metaheuristic method. Industrial & Engineering Chemistry Research. 2023. DOI: 10.1021/acs.iecr.3c03107

[27] Zhang S, Luo Y, Yuan X. A novel stochastic optimization method to efficiently synthesize large-scale nonsharp distillation systems. AICHE Journal. 2021;**67**(9). DOI: 10.1002/AIC.17328

Section 2

Applications

Chapter 4

Introducing Global Optimization vs. Local Optimization in Water Resource Management

Ramadan Abdelaziz

Abstract

There are two types of optimization methods used in Water resources management: global optimization and local optimization. Global optimization algorithms, such as hydroPSO, MCMC, and genetic algorithms (GAs), have been used to optimize large-scale problems with many variables. Local optimization algorithms, such as PEST and UCODE, are best suited for fine-tuning and calibration of models to better fit specific observed data. Optimization techniques have become increasingly significant in water resource management over recent decades, aiming to efficiently allocate resources while minimizing environmental impacts. Two primary approaches dominate the field: global and local optimization. The chapter aims to compare global optimization techniques such as hydroPSO, MCMC, and others to local optimization techniques such as PEST, UCODE, and others in groundwater and surface water modeling. It is a flexible and powerful tool that can handle various types of hydrological models, such as MODFLOW, HYDRUS, and SWAT. The chapter will use different performance measures and case studies to provide a comprehensive comparison of these techniques in water resources management.

Keywords: hydroPSO, MCMC, GA, PEST, UCODE, MODFLOW, MT3DMS, SWAT, HYDRUS

1. Introduction

Water is a valuable and limited resource. To guarantee its sustainable usage, water, a precious and finite resource, needs to be managed carefully [1]. In order to maximize water availability, it involves modeling and optimizing surface water and groundwater systems [2], while also minimizing conflicts between users and preserving the ecological balance [3]. The optimization process is essential in WRM since it enables decision-makers to make educated choices based on precise forecasts and assessments [3].

Water resources modeling is a sophisticated process that uses mathematical models to simulate the dynamic of the water system [4]. Hydraulic conductivity, recharge rate, and storage capacity are among the key parameters that must be estimated for these models [5]. In order to guarantee the correctness of the model outputs, these models must be calibrated [6]. To reduce the discrepancies between the simulated and observed data, the model parameters are adjusted during the calibration process [7]. The crucial phase in the model calibration procedure is the optimization of these parameters [8].

Optimization is the process of identifying the most effective solution among multiple feasible options. In water resource management, this involves determining the optimal distribution of water resources to meet various stakeholders' demands, while accounting for physical and environmental limitations [9]. Optimization is essential in water resource management. The selection of the appropriate optimization method should align with the problem's complexity and the resources at hand [10]. Both global and local optimization techniques come with their own sets of pros and cons, making it crucial to evaluate these aspects carefully when addressing water resource management challenges.

There are two primary optimization methods: global and local. Global optimization seeks the best overall solution within the entire search space, although it can be computationally intensive. It is particularly useful for complex water resource management issues. Local optimization focuses on the best solution within a specific region of the search space, generally operating faster and refining global optimization results. This method is ideal for large search spaces with complex objectives [11, 12].

Selecting the appropriate method depends on the problem's complexity and available computational resources. For extensive systems with many stakeholders and intricate constraints, global optimization ensures the best solution. Conversely, for smaller systems, local optimization offers a quicker and more effective approach [13].

2. Applications of global optimization in water resource management

2.1 Global optimization techniques

Global optimization techniques are frequently employed to determine the ideal parameter set for models with a high level of nonlinearity. By looking for a global minimum in the parameter space, these methods aim to minimize the objective function.

2.1.1 Particle swarm optimization (PSO)

The Particle Swarm Optimization (PSO) algorithm was first introduced by Kennedy and Eberhart in 1995 [14]. It is a population-based stochastic optimization technique inspired by the social behavior of flocks of birds or schools of fish [15]. The PSO algorithm's fundamental premise is to search for the optimal answer in a given search space by imitating the social behavior of a swarm of birds or fish [16]. Each potential solution is represented by a particle in PSO, and the particles move around the search space to discover the best solution based on their personal best and global best positions [17]. It has been widely applied in water resources management, including groundwater management [18], water distribution system optimization [19], and reservoir operation [20].

2.1.1.1 Application of PSO to water resource management problems

They are widely used in different fields of water resource management, such as reservoir operation, rainfall-runoff modeling, water quality modeling, and groundwater modeling. For example, the applications of PSO algorithms in water resource management are:

1. Optimal allocation of urban water resources while maximizing the economic, social, and ecological benefits of the city. Chen et al. [21] proposed a multi-objective optimal allocation model for urban water resources that takes into account the city's economic, social, and environmental benefits, as well as water supply and demand restrictions. The PSO algorithm was used to solve the model and generate optimized water distribution plans for Kunming, China, under various inflow guarantee rates and time horizon scenarios.

2. Hybrid algorithm of PSO and gray wolf optimizer for reservoir operation management. Dahmani and Yebdri [22] suggested a hybrid PSO-gray wolf optimizer (HPSOGWO) method for reservoir operation management that combines the benefits of both algorithms while overcoming their shortcomings. HPSOGWO was utilized to solve two case studies: one with four benchmark functions and another with minimizing the deficit between downstream demand and release from Algeria's Hammam Boughrara reservoir. It demonstrates how it can be integrated and improved with another swarm-based metaheuristic, gray wolf optimizer (GWO), which improves its exploration and exploitation skills.

3. Multi-objective PSO for optimal crop and irrigation water allocation. Yousefi, Maryam, et al. [23] introduced a multi-objective particle swarm optimization (MOPSO) model for the conjunctive use of treated wastewater and groundwater for agricultural irrigation, with the objectives of crop patterning, nitrogen leaching, and aquifer recharging. MOPSO was utilized to solve a case study involving Iran's Varamin irrigation network, and the results were compared to a single-objective PSO model employing an additive weighting method. The method shows that enhancing cropping patterns, water consumption productivity, and aquifer recharge can increase benefits by 7%, 49%, and 20%, respectively, as compared to the current scenario. Furthermore, the method reduces the combined usage of treated wastewater and groundwater by 35%, as well as nitrogen consumption.

4. Parameter estimation of reactive transport models (geochemical modeling). Abdelaziz et al. [24] suggested a method for estimating surface complexation constants from batch measurements by integrating the hydroPSO R optimization tool with the PHREEQC geochemical model. HydroPSO was used to calibrate the PHREEQC model for uranium sorption on quartz at different pH levels, and the results were compared to the well-known parameter estimation (PEST) software. When both software are coupled to PHREEQC and the identical thermodynamic input data is used, hydroPSO can discover more reliable thermodynamic parameter values than PEST. HydroPSO proved that it can give useful post-processing tools for evaluating uncertainty in model parameters and simulated values.

In fact, PSO has been utilized in numerous research pertaining to the management of water resources, including the calibration of rainfall-runoff models [25] and sediment transport modeling [26]. The HydroPSO algorithm was proposed by Abdelaziz and Zambrano-Bigiarini [18] for groundwater modeling. It combines the MODFLOW and MT3DMS model, a commonly used model for groundwater modeling and solute-transport, with the PSO algorithm. The HydroPSO algorithm has been demonstrated to be a powerful global optimization tool for groundwater modeling.

2.1.1.2 Advantages of PSO

PSO has the following advantages: The capacity to handle extremely nonlinear and multimodal objective functions, as well as a large number of decision factors, distinguishes the PSO algorithm from other global optimization approaches. Furthermore, as compared to other global optimization approaches, the PSO algorithm is computationally efficient and requires fewer model evaluations.

2.1.1.3 Limitations of PSO

PSO's key constraints are as follows: PSO can become caught in local optima. Because PSO relies on the global and local optimal positions discovered thus far to lead the swarm, it may converge prematurely to local optima and miss the global optimum. PSO struggles to deal with multimodal challenges. When there are numerous optimal or near-optimal solutions to the objective function, PSO tends to converge on one of them while ignoring the others. This issue is magnified in complicated multimodal scenarios. PSO has a lot of settings that need to be tweaked. PSO performance is influenced by criteria such as swarm size, inertia weight, and cognitive and social acceleration coefficients. Choosing acceptable parameter values for a given problem can be difficult and time-consuming, necessitating trial-and-error experimentation.

2.1.1.4 Conclusion

Finally, PSO has demonstrated potential for use in water modeling and hydrological simulations. PSO can be efficient in optimizing the parameters of water models to increase their performance due to its simplicity, quick convergence, and capacity to find good quality solutions. However, there are numerous limits to PSO that must be noted. For complicated water models with multimodal objective functions, PSO has a propensity to converge prematurely to local optima. This may result in inferior solutions. The performance of PSO is strongly dependent on the parameters that are chosen, which might be challenging to determine for new applications. It has been discovered that combining PSO with other metaheuristics or local search methods can solve some of its shortcomings and increase its effectiveness for water simulation.

Despite the difficulties, PSO remains an attractive optimization approach due to its ease of implementation and efficiency in identifying good solutions with minimal computational work. PSO shows potential as an optimization approach to improve the predictive capability of water models and simulations with correct parameter adjustment and hybridization. To reach the full potential of PSO for water resource modeling applications, future research should focus on overcoming its limitations through algorithm tweaks and hybrid techniques.

The following are the main points stated in this conclusion:

- PSO can be useful for optimizing water models, but it has drawbacks such as local optima and parameter sensitivity.

- Some constraints can be solved, and performance can be improved using hybrid PSO approaches.

- PSO offers promise with proper tuning and modifications, but further research is required to realize its full potential.

2.1.2 Markov chain Monte Carlo (MCMC)

Another method that has gained attention in recent years is Markov Chain Monte Carlo (MCMC) optimization. MCMC is a global optimization technique [27]. It is a stochastic simulation method used to generate samples from a given probability distribution [28]. The method is widely used in Bayesian inference and statistical physics [29]. The basic idea behind MCMC is to construct a Markov chain that has the desired distribution as its equilibrium distribution [30]. The Markov chain is then sampled to generate a sequence of samples that approximate the desired distribution [31]. MCMC is useful for handling complex models and high-dimensional spaces [32].

The algorithm estimates the most probable parameter values by computing the frequency of occurrence of each parameter set in the sequence. The advantages of MCMC optimization include the ability to estimate the uncertainty in the parameter values, the ability to handle non-Gaussian distribution of the model parameters, and the ability to estimate the correlation between parameters [33]. The most used MCMC algorithm is the Metropolis-Hastings algorithm [34, 35].

2.1.2.1 Application of MCMC to water resource management problems

For inverse modeling in a variety of areas related to water resources, including hydrology, hydrogeology, and water quality, MCMC is a flexible and effective method. MCMC has been used in a variety of fields, as shown by several examples:

Rainfall-runoff modeling: Marshall et al. [36] compare four MCMC sampling algorithms in the context of rainfall-runoff modeling to evaluate their performance and efficiency. The four MCMC algorithms compared are a conventional Metropolis-Hastings algorithm with block and single-site updating, an adaptive Metropolis algorithm that adjusts the proposal covariance matrix, a delayed rejection algorithm that uses multiple proposals, and a differential evolution adaptive Metropolis algorithm that uses information from multiple chains.

Groundwater flow modeling: Zhang et al. [37] offer a novel proposal distribution that considerably improves the efficiency of MCMC and is used to determine the parameters and uncertainty of models that mimic how precipitation converts into streamflow. The proposal distribution, which uses the cross-covariance of model parameters, measurements, and model outputs to generate candidate states similar to the analysis stage in the Kalman filter, can significantly enhance MCMC simulation performance. For groundwater models with more than 100 parameters, a speedup of the order of 10–30 times was found.

Modeling of contaminant transport: MCMC can be used to determine the sources, histories, and transport parameters of pollutants in groundwater or surface water systems. Zhou and Tartakovsky [38] proposed an adaptive MCMC method that uses a deep convolutional neural network (CNN) as a surrogate model to replace the computationally expensive transport model and to apply it to the problem of contaminant source identification in groundwater. The novel MCMC uses two simulated examples and one actual case of contaminated source identification in groundwater to show the efficacy of the CNN-based surrogate model. The findings demonstrate that the suggested approach can reach accuracy on par with MCMC using the forward transport model while requiring significantly less computational time.

Modeling of water quality: MCMC can be used to calibrate and assess water quality models that forecast the movement of diseases, metals, and nutrients in aquatic settings. Zheng and Han [39] evaluated the applicability of MCMC in assessing the uncertainty of watershed water quality modeling, using Differential Evolution Adaptive Metropolis (DREAM (ZS)) and SWAT as the representative MCMC algorithm and WWQ model, respectively. It uses MCMC and SWAT to give a case study of nitrate pollution in the Newport Bay watershed. It contrasts various MCMC uncertainty analysis designs and assesses how they affect model performance and management implications. It demonstrates how MCMC approaches, which account for parameter uncertainty and offer probabilistic estimates of model outputs, can enhance the robustness and dependability of WWQ models.

Hydrologic data assimilation: MCMC can be used to update the state variables and parameters of hydrologic models using sequential or combined data assimilation methods that take into account different types of observations. To enhance the estimate of model states and parameters using observations, Abbaszadeh et al. [40] developed a new data assimilation technique based on Particle Filter, Genetic algorithm, and MCMC. It updates the model states and parameters sequentially with observations using a combination of particle filtering, genetic algorithms, and MCMC. Two fictitious case studies and one actual case study of streamflow forecasting are used to apply the methodology. By minimizing the issues with particle degeneracy and sample impoverishment in particle filtering with small ensemble sizes, the suggested method can significantly improve the effectiveness and robustness of data assimilation.

2.1.2.2 Advantages of MCMC

It may provide a detailed representation of the posterior distribution of model parameters and states while accounting for parameter uncertainty and providing probabilistic predictions of model outputs.

It can handle complicated, high-dimensional, or multimodal target distributions that linear or Gaussian approaches may not be able to approximate well.

The cross-covariance of model parameters, measurements, and model outputs can be used to generate candidate states that are more likely to be accepted and enhance simulation efficiency.

2.1.2.3 Limitations of MCMC

One constraint of MCMC for inverse modeling in water resource modeling is that it can be computationally expensive and time-consuming, particularly for complicated and high-dimensional models requiring a significant number of model evaluations.

Another disadvantage of MCMC for inverse modeling of water resources is that it can suffer from particle degeneracy and sample impoverishment issues, which implies that particles tend to converge to a few modes of the target distribution and lose diversity over time.

A further constraint of MCMC for inverse modeling of water resources is that it can be sensitive to several crucial design decisions, such as how many and which model parameters to consider as random, where to fix the non-random model parameters, and which proposal distribution to use.

2.1.2.4 Conclusion

MCMC is a class of methods for sampling from a probability distribution that can be used in the inverse modeling of water resources. By accounting for parameter uncertainty and offering probabilistic estimates of model outputs, MCMC may provide a comprehensive representation of the posterior distribution of model parameters and states. MCMC can also handle complex, high-dimensional, or multimodal target distributions that linear or Gaussian approaches may not be able to approximate well. The cross-covariance of model parameters, measurements, and model outputs can be used by MCMC to generate candidate states that are more likely to be accepted and enhance simulation efficiency.

However, MCMC has significant drawbacks when it comes to inverse modeling for water resource modeling. MCMC can be computationally intensive and time-consuming, particularly for complicated and high-dimensional problems.

2.1.3 Genetic algorithm (GA)

GA is a global optimization technique that replicates natural selection [41], in which the algorithm seeks the optimum solution by picking the fittest individuals from a population and reproducing them with modifications [42]. To develop novel solutions and improve the fitness of the population, the algorithm employs a series of operators such as crossover, mutation, and selection [43]. The GA approach has been used successfully in a variety of hydrological applications, such as groundwater management, parameter estimation, and sensitivity analysis.

2.1.3.1 Application of GA to water resource management problems

Some applications of genetic algorithms to water resource management problems include:

Groundwater model calibration: Calibration of groundwater flow and transport models—Genetic algorithms have been used to calibrate parameters in groundwater flow and transport models. They can explore huge parameter spaces efficiently to find parameter settings that maximize the match between simulated and observed data. Madsen and Perry [44] illustrated how GA can be utilized in a consulting setting to efficiently calibrate groundwater models. GAs were utilized to calibrate model parameters in two groundwater modeling case studies. It demonstrates how GAs can be used to estimate parameters on a MODFLOW model of a New England river basin. The net groundwater flow into the river was employed as the calibration objective. The GA was instructed to act on four model input parameters: recharge, river conductance, and two general head boundaries. The calibrated GA models demonstrated enhanced prediction capabilities, which were beneficial for understanding groundwater flow patterns and developing management strategies.

Reservoir operation optimization: Reservoir operation optimization entails using genetic algorithms to optimize reservoir operating policies and norms in order to maximize water supply, hydropower generation, flood control, and environmental advantages. Reddy and Kumar [45] proposed a GA for identifying optimal reservoir operating plans that maximize benefits across several goals. To optimize reservoir functioning, a multi-objective evolutionary approach based on non-dominated sorting genetic algorithm II (NSGA-II) is applied. To evaluate different operating strategies, the reservoir operation simulation model is built and coupled with the optimization model. The multi-objective evolutionary algorithm was successful in determining a set of optimal reservoir operating strategies that represented the best trade-offs between the various objectives.

Irrigation scheduling: Genetic algorithms have been used to establish ideal irrigation schedules for crops that optimize production while consuming the least amount of water. For this purpose, crop growth models are frequently combined with GAs. Lee et al. [46] created the Irrigation Schedule and Management (ISM) model, which is based on a simulation-optimization framework (Soil Water Atmosphere Plant (SWAP) model with Genetic Algorithm (GA)) and aims to increase the sustainability of optimal crop production under agricultural drought. GAs combined with crop simulation models provide an efficient method for optimizing irrigation scheduling in order to balance yield goals with constrained water availability. The model determines the best combination of Irrigation Water Amount (IWA) and Irrigation Interval (II) by changing Water Stress (WS) in response to environmental circumstances (weather, soils, crops, and bottom boundary conditions) during the course of a growing season. GAs provide an alternative optimization strategy that may be better suited to dealing with the complexity and uncertainty inherent in irrigation scheduling problems, particularly when water is scarce.

Pump scheduling optimization: To reduce pumping costs while satisfying demand needs, genetic algorithms have been employed to optimize pumping schedules in water distribution networks. Candelieri et al. [47] applied GA to the problem of pump schedule optimization in Water Distribution Networks. GAs in conjunction with hydraulic models provide an efficient method for optimizing pump operations in order to reduce energy expenditures while maintaining appropriate service. In place of Bayesian optimization, GAs could be used to identify optimal pump scheduling and control that reduces energy expenditures while meeting pressure and flow requirements. The EPANET simulator was used to assess the energy cost and hydraulic feasibility of a pump schedule.

Water quality management: Genetic algorithms have been used to discover the best locations and sizes of best management practices in order to improve water quality in watersheds and bodies as cost-effectively as possible. Ding et al. [48] created a model that combines a neural network with a genetic algorithm (NN-GA) to forecast river water quality. The technique accurately predicted river water quality characteristics such as pH, dissolved oxygen, biochemical oxygen demand, and chemical oxygen demand. When compared to an NN model without GA optimization, the GA proved effective in improving the neural network design and weights, resulting in lower prediction errors.

Water allocation optimization: Genetic algorithms have been used to optimize water allocation decisions among various user groups and industries in order to accomplish many economic, equity, and environmental sustainability goals. Zhang et al. [49] created a multi-objective optimal water resource allocation model that takes into account intergenerational equity and is based on the non-dominated sorting genetic algorithm II (NSGA-II). The strategy strives to ensure sustainable development of society and economy by balancing water resource consumption among present and future generations. The model was used to a case study of the Huaihe River Basin in China, where the NSGA-2 algorithm was utilized to develop a set of Pareto optimal solutions that reflect the trade-offs between the objectives. It also employs a fuzzy membership function to choose the best solution from the Pareto set and takes into account four objectives: maximizing social benefit, maximizing economic gain, minimizing water shortage ratio, and maximizing discount value. The technique takes into account a variety of restrictions, including water balance, water demand, water supply, and water quality.

2.1.3.2 Advantages of GA

GA is a heuristic search and optimization technique that simulates the natural evolution process. GA can address complex and nonlinear issues that traditional approaches find difficult to solve. GA is capable of dealing with multiple objectives, discrete/continuous variables, noisy settings, and parallel processing. Unlike other methods, GA can find near-optimal solutions with fewer evaluations. To address distinct difficulties, GA can employ a variety of representations and operators. The following are the key benefits in terms of simplicity, adaptability, and efficient search:

Ease of use: GAs require little information about the problem being optimized.

Adaptability: Over generations, GAs "evolve" parameter sets that provide a better fit to observed data, simulating natural adaptation.

Efficient search: GAs can search large and complex parameter spaces efficiently to identify optimal parameter sets. This can be used to calibrate sophisticated hydrological and hydrogeological models.

Global optimization: Unlike gradient-based approaches, GAs tend to identify the global optimal solution rather than getting stuck in local optima.

2.1.3.3 Limitations of GA

GA implementation remains an art that necessitates subject knowledge and experience. GA requires less information about the problem, but it might be challenging to define an objective function and get the representation and operators right. GA is computationally intensive and time-consuming, particularly for large-scale issues. The global optimality or practicality of the solutions cannot be guaranteed by GA. The following are the key restrictions based on convergence, parameter selection, and premature convergence:

- Slow convergence: GAs can take a long time to find the best solution, especially for difficult problems with many parameters.

- Many parameters: The performance of GAs is dependent on appropriately selecting factors such as population size, mutation rate, and so on. This will involve some trial and error.

- Premature convergence: Premature convergence, local optima, and loss of diversity may occur in GA.

2.1.3.4 Conclusion

Groundwater modeling, reservoir operation, water resource allocation, water quality prediction, and other hydrological and hydrogeological problems can all benefit from the application of GA. GA can handle the uncertainty, complexity, and nonlinearity of hydrological and hydrogeological systems while still providing a flexible and resilient solution. To improve the performance and accuracy of the solutions, GA can be integrated with other approaches such as neural networks, principal component analysis, Bayesian optimization, and so on. The main point is as follows:

- In general, GAs are successful at calibrating complex hydrological and hydrogeology models.
- When compared to manually calibrated models, GA-calibrated models exhibit better fits to observed data and improved predictive power.
- With proper use and enhancement, GAs continue to be a valuable modeling approach.
- GA-calibrated models can provide valuable insights and help to build effective water management plans.

In conclusion, while GAs have significant drawbacks, when utilized correctly, they provide an effective method for calibrating model parameters. Calibrated models have better performance and predictive power, which can help with comprehension and decision-making.

3. Applications of local optimization in water resource management

3.1 Local optimization techniques

Local optimization techniques are used to find the optimal solution within a local region of the parameter space. These techniques are particularly useful when the objective function is multimodal or when the search space is restricted due to computational limitations.

3.1.1 Parameter ESTimation (PEST)

PEST (Parameter ESTimation) is a frequently used software tool that improves hydrological model parameter values by reducing the discrepancy between observed and simulated values [50]. PEST is a widely used local optimization software for parameter estimates in hydrology and water resource management, with a focus on groundwater model calibration [51]. Doherty created PEST in 1994, and it has been upgraded since then. The software is compatible with a variety of models, including MODFLOW, HSPF, and SWMM [52]. The software employs a weighted least-squares method that allows for the input of many sorts of data, such as flow rates, water levels, and water quality [53]. Furthermore, the software supports a wide range of models, including finite-difference, finite-element, and finite-volume [54].

PEST iteratively estimates the model's parameters using a Gauss-Newton technique and can handle many parameters [55]. To update the parameters, the technique uses the sensitivity matrix, which is the partial derivative of the output variables with respect to the model parameters [56]. PEST has been utilized in a variety of applications, including groundwater and surface water management, as well as contaminant transport modeling [24].

PEST has been used in numerous studies related to groundwater modeling, such as groundwater recharge estimation [57], modeling of solute transport [58], and optimization of groundwater management strategies [51].

3.1.1.1 Application of PEST to water resource management problems

The PEST algorithm is commonly used for parameter estimation in the fields of hydrology and hydrogeology. Some of the main applications of PEST are:

Groundwater model calibration: PEST is frequently used to calibrate groundwater flow models by determining model parameters such as hydraulic conductivity, recharge rates, and so on that provide the best match between simulated and observed ground heads and/or flows. González-Quirós and Comte [57] show how the PEST method is still being used and developed for calibrating a groundwater flow model using head observations. The method describes how to calibrate groundwater models and analyze uncertainty using the PEST algorithm and the finite-element software COMSOL Multiphysics. By incorporating geophysics into groundwater modeling, the researchers hoped to overcome the limits of standard calibration approaches. It describes the complete integration of PEST and COMSOL, which was accomplished by creating a Java interface that allows the two software packages to communicate in real time. The integrated approach enhanced the accuracy of groundwater model calibration and uncertainty analysis, according to the study.

Surface water model calibration: PEST can also be used to calibrate surface water models [59], such as watershed models, by calculating parameters such as runoff coefficients, infiltration rates, and so on. It employs streamflow and soil moisture data to evaluate the uncertainty of a rainfall-runoff model and uses a hydrologic model alongside climatic projections to gauge the impact of climate change on water supplies [60, 61]. Kim et al. [62] investigated several approaches to upscale nine runoff-related parameters from a series of small monitored plots to the field scale using the KINEROS2/STWIR model and the PEST algorithm for calibration of surface water models and assessing the uncertainty of a rainfall-runoff model. The research presents a case study of rainfall-runoff modeling on a 36-plot field utilizing four seasonal rainfall models. The PEST algorithm was found to be effective in calibrating model parameters and minimizing the uncertainty of model predictions in the study.

Recharge estimation: PEST can be used to estimate aquifer recharge rates by reducing the difference between simulated and observed groundwater levels. Using spring discharge data, estimate the recharge rates and aquifer parameters of a karstic aquifer. Siade et al. [63] calculated natural recharge and its uncertainty for the Antelope Valley groundwater basin using a PEST-calibrated numerical model of groundwater flow and land subsidence. It gives a case study of groundwater management in Antelope Valley, where natural recharge is important in the court's water allocation decision. It uses the MODFLOW-2000 and SUB-WT programs to create a transient model of groundwater flow and land subsidence. It calibrates the model with water-level and subsidence data from PEST. It uses Null-Space Monte Carlo and Pareto trade-off approaches to assess the uncertainty of natural recharging. It also explores the consequences of estimating natural recharge for groundwater management. The calibrated estimate of natural recharge was 36 gigaliters per year (GL/year), which was significantly lower than the court's value (74 GL/year). According to the uncertainty analyses, the total natural recharge is most likely between 34.5 and 54.3 GL/year.

Streamflow simulation: PEST can estimate model parameters in watershed models to provide the best fit between simulated and observed streamflow data at various sites. Using a basic rainfall-runoff model with ten parameters, Goegebeur et al. [64] investigated the performance of the PEST algorithm and the Extended Kalman Filter (EKF) for hydrologic data assimilation. It compares the PEST and EKF approaches for hydrologic data assimilation using a synthetic-data experiment and two case studies. It demonstrates that PEST is quite similar to EKF but has certain changes due to the algorithms' derivation. It also suggests a simple alternative to the Marquardt algorithm for improving PEST's convergence. To account for the trustworthiness of the observations, the study assigned different weights to them. Observations of streamflow during high flow times are given lower weights.

Transport model calibration: PEST has also been used to calibrate groundwater transport models by calculating parameters such as dispersion coefficients, decay rates, and so on. PEST is employed to calibrate groundwater transport models by determining parameters like dispersion coefficients and decay rates. It uses hydraulic conductivity and contaminant concentration data to optimize groundwater remediation system designs [65]. Skahill et al. [58] presented a case study of model calibration using MICUT and PEST for three environmental models (HSPF, FASST, and GSSHA). It explains some of the features of the Model Independent Calibration and Uncertainty Analysis Toolbox (MICUT), which was created to support the popular PEST model independent interface. Jesudhas et al. [66] developed a contaminant transport model using the total dissolved solids (TDS) concentration of groundwater measured in the upper Kodaganar basin, located in Dindigul district of Tamil Nadu, India, where groundwater is being contaminated by tannery wastewater that is discharged into the nearby Sengulam Lake. The calibrated groundwater flow and contaminant transport model accurately predicted observed groundwater levels and contamination concentrations.

3.1.1.2 Advantages of PEST

- Automated and effective parameter calibration: PEST's optimization techniques enable it to effectively calibrate a number of parameters simultaneously. In comparison to manual trial-and-error calibration, this is a substantial advantage.

- Measures the goodness of fit objectively: PEST measures how well the model fits the observed data using objective functions and error metrics. This offers a neutral foundation for contrasting various calibration runs.

- Flexibility: A variety of models and parameters can be calibrated using PEST. It is independent of the model.

- Using pilot points and other parameterization tools, PEST provides complex and adaptable parameterization of groundwater models.

- Tikhonov, subspace, and hybrid schemes are just a few of the mathematical regularization techniques that PEST offers, and they can all help groundwater model calibration be more accurate and effective.

3.1.1.3 Limitations of PEST

- Computationally demanding: PEST necessitates a high number of model runs, which might be prohibitively expensive for sophisticated large-scale models.

- Prone to being stuck in local optima: PEST's optimization techniques might sometimes converge to local optima rather than the global optimum.

- Requires suitable beginning parameter ranges: PEST performs best when given reasonable initial parameter ranges near to ideal values.

3.1.1.4 Conclusion

In general, PEST is an effective parameter calibration method that can improve the prediction performance of hydrological and hydrogeological models. Some of PEST's limitations have been addressed through adjustments and upgrades, increasing its effectiveness and efficiency. Surrogate model-based sequential calibration methodologies show promise in making PEST calibration feasible for large-scale models. It has been demonstrated that PEST-calibrated models improve the simulation of processes such as groundwater flow, pollutant transport, and streamflow.

In conclusion, while PEST has significant limitations, it remains a powerful tool for parameter estimate and model calibration in the disciplines of water resource management with appropriate application and enhancements.

3.1.2 UCODE

UCODE is an acronym that stands for Uncertainty analysis and CODE optimization [67]. Another well-known local optimization software for parameter estimate in hydrology and water resource management is UCODE [56]. It is a local optimization technique developed for model calibration in environmental systems [68]. UCODE is a model parameter handling approach that is based on the Shuffled Complex Evolution (SCE) algorithm [69]. UCODE is very beneficial for tasks that have a high level of uncertainty and variability [70]. Depending on the needs of the user, the software employs a set of algorithms that enable both local and global optimization [63]. UCODE can also simulate parameter sets using Monte Carlo simulation to perform uncertainty analysis [71]. UCODE, like PEST, can support a variety of models, including MODFLOW, MT3DMs, PHREEQC, and others [72].

3.1.2.1 Application of UCODE to water resource management problems

UCODE has been used in various studies related to water resources management, such as:

Groundwater model calibration: UCODE is often used to calibrate groundwater flow and transport models by estimating parameters such as hydraulic conductivity, recharge rates, boundary fluxes, and so on. This increases the models' forecasting potential for use in management research. Sanford et al. [73] calibrate a model using UCODE to assess groundwater flow patterns and the effects of groundwater pumping on the system. The Middle Rio Grande Basin in central New Mexico, USA, is the study area, and the goal is to quantify the forecast uncertainty for recharge sites. The calibrated model accurately predicted groundwater levels, flow patterns, and tracer concentrations. Pumping has also caused river seepage and streamflow depletion, affecting surface water supplies.

Streamflow modeling: UCODE can be used to calibrate watershed and surface water models by calculating parameters such as infiltration rates, surface runoff coefficients, base flow coefficients, and so on. This aids in the improvement of streamflow simulations. Geza et al. [74] used a Watershed Analysis Risk Management Framework (WARMF) model with a nonlinear parameter estimation code (UCODE) to quantify the predictive uncertainty for stream flow in a mountain-watershed in Colorado. WARMF model of the Turkey Creek Watershed was affected by land use, climate, and soil conditions. The calibrated model reasonably closely reflects the observed stream flow. For regional flow modeling, the effects of land use changes on stream flow and interannual recharge variability must be considered.

Recharge estimation: UCODE can be used to estimate aquifer recharge rates by reducing the difference between simulated and observed groundwater levels. The estimated recharge rates are useful for management studies. Dripps et al. [75] employed a nonlinear parameter estimate code (UCODE) in conjunction with an analytic element (AE) code (GFLOW) to quantify the geographical and temporal distribution of recharge using measured base flows as calibration targets. It employs an AE flow model of an undeveloped watershed in northern Wisconsin, where vegetation types and precipitation influence recharging. It concludes that the calibrated model fits the observed data set well and that hydraulic conductivity and immobile porosity are major factors in recharge variability.

Transport modeling: UCODE has been used to calibrate groundwater pollutant transport models by calculating parameters such as dispersion coefficients, decay rates, and so on. This contributes to more accurate forecasts of contamination migration and remediation timelines. Abdelaziz and Merkel [72] used UCODE, MODFLOW, and MT3DMS to calibrate a conservative solute transport model based on a double-porosity approach using inverse modeling. Several statistics are used to do sensitivity analysis and to investigate the interplay of model parameters. It analyzes the impact of hydraulic conductivity and immobile porosity on the transport process, as well as the model parameter uncertainty.

Sensitivity analysis: Sensitivity analysis is made possible by UCODE's Monte Carlo and Latin hypercube sampling capabilities, which enable thorough sensitivity analyses of models for groundwater and surface water. This identifies significant parameters that influence model outcomes and management objectives. Starn and Bagtzoglou [76] propose an automated calibrated Monte Carlo (CMC) approach for defining recharge and capture zones for wells. The CMCs method calibrates a groundwater model of a mountain-watershed in Connecticut, where recharge regions are modified by land use and geology, and then generates an ensemble of equiprobable parameter choices that honor the calibration data. These parameter sets are then employed in Monte Carlo simulation to produce probabilistic recharge and capture zone delineations that take parameter uncertainty into account.

Optimization of water allocation policies: UCODE techniques like model calibration, uncertainty analysis, incorporation of multiple objectives, global optimization, and evaluation of alternatives all offer ways to systematically optimize water allocation policies to achieve desired economic, social, and environmental outcomes in a robust manner. Nikoo [77] calibrated the model parameters using historical soil moisture and crop yield data as calibration objectives. UCODE calibrated a nonlinear regression model for the irrigation simulation model (ISM) utilizing historical soil moisture and crop yield data as calibration objectives. The parameter covariance matrix and correlation coefficients for the ISM model were developed by UCODE and are utilized as inputs for the Genetic algorithm (GA) optimization model. The strategy aided in the allocation of irrigation water and the timing of irrigation.

3.1.2.2 Advantages of UCODE

UCODE has the following advantages:

- Automated calibration: Using built-in optimization methods, UCODE can efficiently calibrate several parameters at once. When compared to manual calibration, this saves time.

- Inverse modeling technique: To discover the parameters that give the best fit between simulated and observed data, UCODE employs an inverse modeling approach.

- Global and local optimization: UCODE provides global and local optimization capabilities to find the best global or local solutions.

- High adaptability: UCODE can calibrate a wide variety of hydrological and hydrogeological models. It is not model-dependent.

By employing a standard input and output format, UCODE can ease the coupling of hydrological models with other optimization approaches such as genetic algorithms.

3.1.2.3 Limitations of UCODE

UCODE's limitations are:

- Require appropriate initial parameter ranges: UCODE performs best when given initial parameter estimations that are reasonable and not too far off from the ideal values.

- Prone to being stuck in local optima: Local optimization methods may occasionally converge to local optima rather than the global optimum.

- Computationally demanding: Using UCODE to calibrate complex models can necessitate a significant number of model runs.

3.1.2.4 Conclusion

In general, UCODE is a powerful parameter calibration tool that can greatly improve the performance of hydrological and hydrogeological models. It has been demonstrated that calibrated models utilizing UCODE can better simulate processes such as groundwater flow, transport, and streamflow. UCODE can propagate parameter uncertainty through models to quantify uncertainty in model outputs. This provides a more robust basis for water resource management decisions under uncertainty. UCODE can be combined with other optimization techniques to improve its efficacy and effectiveness in addressing challenging water allocation issues.

To summarize, UCODE improves the knowledge and prediction power of models used for a variety of water resource management applications and decision-making by enabling quick calibration and analysis of groundwater, surface water, and pollutant transport models. When properly applied and refined, UCODE continues to be a valuable tool for model calibration in hydrology and hydrogeology applications. Models calibrated using UCODE can provide useful insights and help to build effective water management plans. So, while UCODE has significant limitations, when used correctly, it provides an efficient and automated method for calibrating model parameters. Calibrated models have better performance and predictive power, which can help with comprehension and decision-making.

4. Comparison of the effectiveness of global and local optimization from mathematical point of view

These mathematical expressions aid in the understanding of the fundamental optimization principles behind techniques such as ACO, MCMC, PSO, GA, PEST, and UCODE—whether global or local optimization. Specific formulations then differ depending on the algorithms and applications.

The basic mathematical equations for various typical global and local optimization techniques are provided here:

4.1 ACO

A global optimization technique inspired by ant foraging behavior is defined as [78]:

$$P_{i,j} = \frac{\left(\tau_{i,j}^{\alpha}\right)\left(\eta_{i,j}^{\beta}\right)}{\sum\left(\tau_{i,j}^{\alpha}\right)\left(\eta_{i,j}^{\beta}\right)} \tag{1}$$

where

$P_{i,j}$ is the probability of choosing path j from i.

$\tau_{i,j}^{\alpha}$ is the pheromone trail concentration.

$\eta_{i,j}^{\beta}$ is the heuristic value (inverse of distance).

α and β are parameters to control influences for $\tau_{i,j}^{\alpha}$ and $\eta_{i,j}^{\beta}$.

Local Pheromone Update:

After an ant traverses an edge (i, j), the pheromone level on that edge is locally updated as follows:

$$\tau_{i,j} = (1-\rho)\tau_{i,j} + \rho\Delta\tau_{i,j} \tag{2}$$

where ρ is the pheromone evaporation rate, and $\Delta\tau_{i,j}$ represents the amount of pheromone deposited by the ant.

Global Pheromone Update:

After all ants complete their tours, the pheromone level on the edges is globally updated based on the quality of the solutions found. The update is performed as follows [79]:

$$\tau_{i,j} = (1-\rho)\tau_{i,j} + \sum\left(\Delta\tau_{i,j}^{k}\right) \tag{3}$$

where $\Delta\tau_{ij}^{k}$ represents the amount of pheromone deposited by ant k on edge (i, j).

4.2 MCMC

MCMC is a computational technique used for sampling from probability distributions. The basic mathematical equations for MCMC include:

Proposal Distribution:

q(x|y) is the probability density function (PDF) of proposing a new state x given the current state y [80].

Acceptance Probability:

$$\alpha(x|y) = min\,(1, P(x)/P(y)) * (q(y|x)/q(x|y)) \tag{4}$$

where P(x) and P(y) are the probabilities of states x and y, respectively [81].

4.3 PSO

Bird flocking behavior inspired a global optimization technique. The basic mathematical equations for PSO include [82]:

Velocity Update:

$$v_i(t+1) = \omega v_i(t) + c_1 r_1 \left(pbest_i - x_i(t)\right) + c_2 r_2 (gbest - x_i(t)) \tag{5}$$

where

$v_i(t)$ is the velocity of particle i

ω is the inertia weight

c_1 and c_2 are constants

r_1 and r_2 are random numbers between 0 and 1

$pbest_i$ is the best previous position of particle i

$gbest$ is the best position of the group

Position Update:

$$x_i(t+1) = x_i(t) + v_i(t+1) \tag{6}$$

Where $x_i(t)$ is the current position of particle i [83].

4.4 GA

A technique for global optimization called the genetic algorithm was developed as a result of natural evolution [84]. Using crossover and mutation operators, the fundamental equation for generating a new population of solutions from the existing population is:

$$x' = crossover(x, x') + mutation(x') \tag{7}$$

where x and x' are two parent solutions chosen from the current population based on their fitness values, $crossover(x, x')$ is a function that combines parts of x and x' to form a new solution, and $mutation(x')$ is a function that alters certain aspects of x' at random to introduce diversity.

4.5 Pest

Local optimization technique that minimizes an objective function.

$$Minimize\ \varphi = (m - o)T(m - o) + \alpha\varphi \mathrm{P} \tag{8}$$

where
φ is the objective function
m is the model response vector
o is the observation vector
α is a trade-off coefficient
φP is a penalty function

4.6 UCODE

UCODE is divided into three parts: parameter estimation, sensitivity analysis, and uncertainty analysis [85].

Parameter estimation is the process of determining the ideal model parameter values that minimize an objective function that assesses the difference between model outputs and observations [86]. The parameter estimation problem is stated as follows:

$$p^{min\,\Phi(p)} = p^{min} \sum_{i=1}^{N} h_i(p) - o_i)^2 \tag{9}$$

where p denotes the model parameter vector, Φ(p) the objective function, N the number of observations, wi the weight of observation i, $h_i(p)$ the model output corresponding to observation i, and oi the observed value of observation i.

The technique of determining how sensitive model outputs are to changes in model parameters or inputs is known as sensitivity analysis [68]. The following is an expression for the sensitivity analysis problem:

$$s_{ij} = \frac{\partial h_i}{\partial p_j} \tag{10}$$

where s_{ij} denotes the sensitivity coefficient of observation i to parameter j and $\partial p_i/\partial h_j$ denotes the partial derivative of model output h_i to parameter p_j.

The technique of assessing the uncertainty in model parameters and outputs due to uncertainty in observations and previous information is known as uncertainty analysis [67]. The problem of uncertainty analysis can be expressed as follows:

$$p = p^{\wedge} + u \tag{11}$$

where p is the vector of model parameters, u is the vector of parameter uncertainty, and p^ is the vector of optimal parameter values determined by parameter estimation. Various methods, such as linear error propagation, Monte Carlo simulation, and Bayesian inference, can be used to estimate parameter uncertainty.

5. Recommendations for choosing between global and local optimization in water resource management

The optimization approach used should be determined by the characteristics of the water resource problem, such as model type, data, objective function, restrictions, and uncertainty. When deciding between global and local optimization strategies for water resource management applications, consider factors such as issue complexity, parameter estimates, objective function form, computational needs, desired optimality, and associated trade-offs. Hybrid techniques may potentially provide a useful solution.

Global optimization approaches, such as GA, ACO, and PSO, and MCMC are well suited for solving complex and nonlinear water resource problems with many objectives, constraints, and uncertainties.

Following are some suggestions for deciding between global and local optimization strategies in water resource management applications:

Problem complexity: Global optimization techniques are generally advised for more difficult problems with nonlinear and non-convex parameter spaces. For simpler linear or convex problems, local optimization may be adequate.

Initial parameter estimates: If initial parameter estimations are far from ideal, global optimization is more likely to find the genuine optimum. Local optimization works better when the initial values are reasonable.

Objective function: If the objective function has various local optima, global optimization approaches are required to identify the global optimum. Local optimization can become stuck in local optima.

Computational demand: Global optimization approaches examine the entire parameter space, they are more computationally demanding. Local optimization is faster, but it may converge faster.

Desired optimality: Global optimization should be employed if the genuine global optimum is required to give the most optimal water management solutions. Local optimization finds a suboptimal local optimum.

Trade-off analysis: When deciding between global and local optimization for a specific problem, a trade-off analysis between optimality, computational needs, and the certainty of the optimum obtained can help.

Hybrid methods: Hybrid strategies that begin with local optimization to narrow in on an optimum and then move to global optimization to confirm the global optimum have also been investigated.

Author details

Ramadan Abdelaziz
gbm Gesellschaft für Baugeologie und-meßtechnik mbH Baugrundinstitut Society for Building Geology and Measurement Technology Building Ground Institute, Ettlingen, Germany

*Address all correspondence to: ramawaad@gmail.com

References

[1] Water Resources Management. World Bank. Available from: https://www.worldbank.org/en/topic/waterresourcesmanagement [Accessed: April 3, 2023]

[2] Hart S. Water Resources Management: Principles and Practice. New York, United States: Syrawood Publishing House; 2022. [Online]. Available from: https://books.google.de/books?id=pmO1zgEACAAJ

[3] Loucks DP, van Beek E, Stedinger JR, Dijkman JPM, Villars MT. Water Resources Systems Planning and Management: An Introduction to Methods, Models and Applications. Paris: UNESCO; 2005. [Online]. Available from: https://ecommons.cornell.edu/handle/1813/2804 [Accessed: April 2, 2023]

[4] Mujumdar P. Water Resources Systems: Modeling Techniques and Analysis. Boca Raton, FL, USA: CRC Press; 2014.

[5] Sophocleous M. Groundwater recharge and sustainability in the High Plains aquifer in Kansas, USA. Hydrogeology Journal. 2005;**13**:351-365

[6] Beven KJ. Rainfall-Runoff Modelling: The Primer. Hoboken, New Jersey, United States: John Wiley & Sons; 2011

[7] Duan Q, Sorooshian S, Gupta V. Effective and efficient global optimization for conceptual rainfall-runoff models. Water Resources Research. 1992;**28**(4):1015-1031

[8] Moriasi DN, Arnold JG, Van Liew MW, Bingner RL, Harmel RD, Veith TL. Model evaluation guidelines for systematic quantification of accuracy in watershed simulations. Transactions of the ASABE. 2007;**50**(3):885-900

[9] Boyd S, Boyd SP, Vandenberghe L. Convex Optimization. Cambridge, United Kingdom: Springer Science & Business Media; 2004

[10] Giuliani M, Castelletti A. Assessing the value of cooperation and information exchange in large water resources systems by agent-based optimization. Water Resources Research. 2013;**49**(7):3912-3926. DOI: 10.1002/wrcr.20287

[11] Pardalos PM, Romeijn HE. Handbook of Global Optimization: Volume 2. Vol. 62. New York, United States: Springer Science & Business Media; 2013

[12] Horst R, Pardalos PM. Handbook of Global Optimization. Vol. 2. New York, United States: Springer Science & Business Media; 2013

[13] Maier HR et al. Evolutionary algorithms and other metaheuristics in water resources: Current status, research challenges and future directions. Environmental Modelling & Software. 2014;**62**:271-299

[14] Kennedy J, Eberhart R. Particle swarm optimization. In: Presented at the Proceedings of ICNN'95-International Conference on Neural Networks. Piscataway, NJ, USA: IEEE; 1995. pp. 1942-1948

[15] Duan H, Ma G, Luo D. Optimal formation reconfiguration control of multiple UCAVs using improved particle swarm optimization. Journal of Bionic Engineering. 2008;**5**(4):340-347

[16] Wang D, Tan D, Liu L. Particle swarm optimization algorithm: An overview. Soft Computing. 2018;**22**:387-408

[17] Juneja M, Nagar S. Particle swarm optimization algorithm and its parameters: A review. In: Presented at the 2016 International Conference on Control, Computing, Communication and Materials (ICCCCM). Piscataway, New Jersey, United States: IEEE; 2016. pp. 1-5

[18] Abdelaziz R, Zambrano-Bigiarini M. Particle swarm optimization for inverse modeling of solute transport in fractured gneiss aquifer. Journal of Contaminant Hydrology. 2014;**164**:285-298

[19] Patil MB, Naidu MN, Vasan A, Varma MR. Water distribution system design using multi-objective particle swarm optimisation. Sādhanā. 2020;**45**: 1-15

[20] Afshar M. Large scale reservoir operation by constrained particle swarm optimization algorithms. Journal of Hydro-environment Research. 2012;**6** (1):75-87

[21] Chen J, Yu C, Cai M, Wang H, Zhou P. Multi-objective optimal allocation of urban water resources while considering conflict resolution based on the PSO algorithm: A case study of Kunming, China. Sustainability. 2020;**12**(4):1337

[22] Dahmani S, Yebdri D. Hybrid algorithm of particle swarm optimization and grey wolf optimizer for reservoir operation management. Water Resources Management. 2020;**34**(15): 4545-4560

[23] Yousefi M, Banihabib ME, Soltani J, Roozbahani A. Multi-objective particle swarm optimization model for conjunctive use of treated wastewater and groundwater. Agricultural Water Management. 2018;**208**:224-231

[24] Abdelaziz R, Merkel BJ, Zambrano-Bigiarini M, Nair S. Particle swarm optimization for the estimation of surface complexation constants with the geochemical model PHREEQC-3.1. 2. Geoscientific Model Development. 2019; **12**(1):167-177

[25] Zhang C, Sun YY. Automatic calibration of conceptual rainfall-runoff models. In: Presented at the Applied Mechanics and Materials, Berlin, Germany: Trans Tech Publications Ltd; 2013. pp. 2185-2189

[26] Mohammadi B, Guan Y, Moazenzadeh R, Safari MJS. Implementation of hybrid particle swarm optimization-differential evolution algorithms coupled with multi-layer perceptron for suspended sediment load estimation. Catena. 2021; **198**:105024

[27] Sen MK, Stoffa PL. Global Optimization Methods in Geophysical Inversion. Cambridge, United Kingdom: Springer Science & Business Media; 2013

[28] Karras C, Karras A, Avlonitis M, Sioutas S. An overview of MCMC methods: From theory to applications. In: Presented at the Artificial Intelligence Applications and Innovations. AIAI 2022 IFIP WG 12.5 International Workshops: MHDW 2022, 5G-PINE 2022, AIBMG 2022, ML@ HC 2022, and AIBEI 2022, Hersonissos, Crete, Greece, June 17–20, 2022, Proceedings. New York, United States: Springer; 2022. pp. 319-332

[29] Ottosen TA. Markov-chain Monte-Carlo A Bayesian approach to statistical mechanics. ArXiv Prepr. ArXiv12066905. 2012

[30] Gilks WR, Richardson S, Spiegelhalter DJ. Introducing Markov chain Monte. In: Markov Chain Monte Carlo in Practice. Vol. 1. London, UK: Chapman & Hall/CRC;1995

[31] Carlo CM. Markov chain Monte Carlo and gibbs sampling. Lecture Notes EEB. 2004;**581**(540):3

[32] Laloy E, Vrugt JA. High-dimensional posterior exploration of hydrologic models using multiple-try DREAM (ZS) and high-performance computing. Water Resources Research. 2012;**48**(1): 1-18

[33] Ford EB. Improving the efficiency of Markov chain Monte Carlo for analyzing the orbits of extrasolar planets. The Astrophysical Journal. 2006;**642**(1):505

[34] Metropolis N, Rosenbluth AW, Rosenbluth MN, Teller AH, Teller E. Equation of state calculations by fast computing machines. The Journal of Chemical Physics. 1953;**21**(6):1087-1092

[35] Hastings WK. Monte Carlo Sampling Methods Using Markov Chains and their Applications. Oxford, UK: Biometrika Trust; 1970

[36] Marshall L, Nott D, Sharma A. A comparative study of Markov chain Monte Carlo methods for conceptual rainfall-runoff modeling. Water Resources Research. 2004;**40**(2):309-319

[37] Zhang J, Vrugt JA, Shi X, Lin G, Wu L, Zeng L. Improving simulation efficiency of MCMC for inverse modeling of hydrologic systems with a Kalman-inspired proposal distribution. Water Resources Research. 2020;**56**(3): e2019WR025474

[38] Zhou Z, Tartakovsky DM. Markov chain Monte Carlo with neural network surrogates: Application to contaminant source identification. Stochastic Environmental Research and Risk Assessment. 2021;**35**:639-651

[39] Zheng Y, Han F. Markov chain Monte Carlo (MCMC) uncertainty analysis for watershed water quality modeling and management. Stochastic Environmental Research and Risk Assessment. 2016;**30**:293-308

[40] Abbaszadeh P, Moradkhani H, Yan H. Enhancing hydrologic data assimilation by evolutionary particle filter and Markov chain Monte Carlo. Advances in Water Resources. 2018;**111**: 192-204

[41] Reeves C, Rowe JE. Genetic Algorithms: Principles and Perspectives: A Guide to GA Theory. Vol. 20. New York, United States: Springer Science & Business Media; 2002

[42] Domashova JV, Emtseva SS, Fail VS, Gridin AS. Selecting an optimal architecture of neural network using genetic algorithm. Procedia Computer Science. 2021;**190**:263-273

[43] Tabassum M, Mathew K. A genetic algorithm analysis towards optimization solutions. International Journal of Digital Information and Wireless Communications. 2014;**4**(1):124-142

[44] Madsen KM, Perry AE. Using genetic algorithms on groundwater modeling problems in a consulting setting. In: Presented at the Proceedings of the Annual International Conference on Soils, Sediments, Water and Energy. Association for Environmental Health and Sciences (AEHS) Foundation (with Curran Associates, Inc. as print-on-demand publisher; 2010. p. 11

[45] Reddy MJ, Kumar DN. Optimal reservoir operation using multi-objective evolutionary algorithm. Water Resources Management. 2006;**20**:861-878

[46] Lee T et al. Development of irrigation schedule and management model for sustaining optimal crop

production under agricultural drought. Paddy and Water Environment. 2023;**21**(1):31-45

[47] Candelieri A, Perego R, Archetti F. Bayesian optimization of pump operations in water distribution systems. Journal of Global Optimization. 2018;**71**:213-235

[48] Ding Y, Cai Y, Sun P, Chen B. The use of combined neural networks and genetic algorithms for prediction of river water quality. Journal of Applied Research and Technology. 2014;**12**(3):493-499

[49] Zhang J, Dong Z, Chen T. Multi-objective optimal allocation of water resources based on the NSGA-2 algorithm while considering intergenerational equity: A case study of the middle and upper reaches of Huaihe River Basin, China. International Journal of Environmental Research and Public Health. 2020;**17**(24):9289

[50] Doherty JE. Methodologies and Software for PEST-Based Model Predictive Uncertainty Analysis. Watermark Numerical Computing, 2010:1-33

[51] Ramadan A, Mahmoud Ibrahim B. Inverse modeling of groundwater flow of Delta Wadi El-Arish. Journal of Water Resource and Protection. 2012;**2012**:432-438

[52] Daniel EB, Camp JV, LeBoeuf EJ, Penrod JR, Dobbins JP, Abkowitz MD. Watershed modeling and its applications: A state-of-the-art review. Open Hydrology Journal. 2011;**5**(1):12

[53] Edet A, Abdelaziz R, Merkel B, Okereke C, Nganje T. Numerical groundwater flow modeling of the coastal plain sand aquifer, Akwa Ibom State, SE Nigeria. Journal of Water Resource and Protection. 2014;**2014**:197

[54] Abdelaziz R. Modeling of Groundwater Flow and Transport in Fractured Aquifers. 2014.

[55] Qin Y, Kavetski D, Kuczera G, McInerney D, Yang T, Guo Y. Can Gauss-Newton algorithms outperform stochastic optimization algorithms when calibrating a highly parameterized hydrological model? A case study using SWAT. Water Resources Research. 2022;**58**(11):e2021WR031532

[56] Hill MC, Tiedeman CR. Effective Groundwater Model Calibration: With Analysis of Data, Sensitivities, Predictions, and Uncertainty. Hoboken, New Jersey, United States: John Wiley & Sons; 2006

[57] González-Quirós A, Comte J-C. Hydrogeophysical model calibration and uncertainty analysis via full integration of PEST/PEST++ and COMSOL. Environmental Modelling & Software. 2021;**145**:105183

[58] Skahill BE, Baggett JS, Frankenstein S, Downer CW. More efficient PEST compatible model independent model calibration. Environmental Modelling & Software. 2009;**24**(4):517-529

[59] Abdelaziz R. Application of ant colony optimization in water resource management. In: Optimization Algorithms-Classics and Recent Advances. London, UK, London, UK: IntechOpen; 2023

[60] Loucks DP, van Beek E. Water Resource Systems Planning and Management: An Introduction to Methods, Models, and Applications. Springer; 2017

[61] McKinney DC, Cai X, Lasdon LS, Watkins DW, Nielsen SG. Integrated

water resources management model for the Rio Grande basin. Regional Environmental Change. 1999

[62] Kim K et al. Rainfall–runoff model parameter estimation and uncertainty evaluation on small plots. Hydrological Processes. 2014;**28**(20):5220-5235

[63] Siade A, Nishikawa T, Martin P. Natural recharge estimation and uncertainty analysis of an adjudicated groundwater basin using a regional-scale flow and subsidence model (Antelope Valley, California, USA). Hydrogeology Journal. 2015;**23**(6):1267

[64] Goegebeur M, Pauwels VR. Improvement of the PEST parameter estimation algorithm through extended Kalman filtering. Journal of Hydrology. 2007;**337**(3–4):436-451

[65] Abdelaziz R, Le HH. MT3DMSP–A parallelized version of the MT3DMS code. Journal of African Earth Sciences. 2014;**100**:1-6

[66] Jesudhas CJ, Chidambaram SM, Jeyakumar RB, Rene ER. Development and application of a contaminant transport model for groundwater remediation and reservoir protection: A case study from India. Environmental Monitoring and Assessment. 2022;**194**(4):257

[67] Poeter EE, Hill MC, Banta ER, Mehl S, Christensen S. UCODE_2005 and six other computer codes for universal sensitivity analysis, calibration, and uncertainty evaluation constructed using the JUPITER API. US Geological Survey, 2328–7055; 2005

[68] Poeter EP, Hill MC. Documentation of UCODE, a Computer Code for Universal Inverse Modeling. Darby, Pennsylvania, United States: Diane Publishing; 1998

[69] Pettway JS. Parameter Estimation Tools for Hydrologic and Hydraulic Simulations. Davis, CA, USA: U.S. Army Corps of Engineers (HEC); 2006.

[70] Hill MC. Methods and guidelines for effective model calibration. In: Hotchkiss RH, Glade M, editors. Building Partnerships. Reston, VA, USA: American Society of Civil Engineers (ASCE); 2000. pp. 1-10

[71] Lu D, Ye M, Hill MC, Poeter EP, Curtis GP. A computer program for uncertainty analysis integrating regression and Bayesian methods. Environmental Modelling & Software. 2014;**60**:45-56

[72] Abdelaziz R, Merkel BJ. Sensitivity analysis of transport modeling in a fractured gneiss aquifer. Journal of African Earth Sciences. 2015;**103**:121-127

[73] Sanford WE, Plummer LN, McAda DP, Bexfield LM, Anderholm SK. Hydrochemical tracers in the middle Rio Grande Basin, USA: 2. Calibration of a groundwater-flow model. Hydrogeology Journal. 2004;**12**:389-407

[74] Geza M, Poeter EP, McCray JE. Quantifying predictive uncertainty for a mountain-watershed model. Journal of Hydrology. 2009;**376**(1–2):170-181

[75] Dripps W, Hunt R, Anderson M. Estimating recharge rates with analytic element models and parameter estimation. Groundwater. 2006;**44**(1): 47-55

[76] Starn JJ, Bagtzoglou AC. Programs for calibration-based Monte Carlo simulation of recharge areas. Groundwater. 2012;**50**(3):472-476

[77] Nikoo MR et al. A coupled water allocation simulation–optimization model to advance agricultural water

management. Arabian Journal of Geosciences. 2022;**15**(5):397

[78] Mosa MA, Hamouda A, Marei M. Ant colony heuristic for user-contributed comments summarization. Knowledge-Based Systems. 2017;**118**: 105-114

[79] Ning J, Zhang Q, Zhang C, Zhang B. A best-path-updating information-guided ant colony optimization algorithm. Information Sciences. 2018; **433**:142-162

[80] Rosenthal JS. Optimal proposal distributions and adaptive MCMC. Handbook of Markov Chain Monte Carlo. 2011;**4**(10):1201

[81] Martino L, Del Olmo VP, Read J. A multi-point Metropolis scheme with generic weight functions. Statistics & Probability Letters. 2012;**82**(7):1445-1453

[82] Helwig S. Particle Swarms for Constrained Optimization Partikelschwarme für Optimierungsprobleme mit Nebenbedingungen [PhD dissertation]. Leipzig, Germany: University of Leipzig; 2010

[83] Mirjalili S, Mirjalili S. Particle swarm optimisation. In: Evolutionary Algorithms and Neural Networks: Theory and Applications. Cham, Switzerland: Springer International Publishing; 2019. pp. 15-31

[84] Houck CR, Joines J, Kay MG. A genetic algorithm for function optimization: A Matlab implementation. NCSU-IE TR. 1995;**95**(09):1-10

[85] Lin G, Engel DW, Eslinger PW. Survey and Evaluate Uncertainty Quantification Methodologies. Richland, WA: Pacific Northwest National Lab (PNNL); 2012

[86] Foglia L, Hill MC, Mehl SW, Burlando P. Sensitivity analysis, calibration, and testing of a distributed hydrological model using error-based weighting and one objective function. Water Resources Research. 2009;**45**(6): 1-18

Chapter 5

Programming Optimisation of Move with Interleaving Transform

Borut Žalik, Aljaž Jeromel, Ivana Kolingerová, Niko Lukač and Blaž Repnik

Abstract

Computer resources and programming solutions have always been utilised with the aim of improving efficiency; consequently, many optimisations have been developed in the past. This chapter first considers some typical characteristic optimisations in computing. Following this, a string transformation technique, Move with Interleaving (MwI), is explained briefly. MwI may become considerably slow when a larger number of elements need to be frequently interleaved in the MwI data structure. The aim of this chapter is to demonstrate how different implementations of this data structure affect CPU time on three different computing platforms: a personal computer running MS Windows and Linux, and a Raspberry Pi 2 using Linux. MwI was implemented in three ways: using a standard vector, a lookup table, and a dynamic linked list supported by a pool of released records. All three approaches were implemented in C++ and Java. Thirty-two 8-bit-plane greyscale raster images were used for experiments, with these images varying in content and resolution. The results show that the implementation using the dynamic linked list is the most optimised and outperforms significantly the implementation based on the lookup table and even more so the implementation using the standard vector. The most dramatic differences were observed on the less powerful Raspberry Pi for the implementation in Java.

Keywords: algorithms, data structures, string transformation, C++, Java

1. Introduction

This chapter considers different implementations of Move with Interleaving (MwI) string transformation technique with the aim to optimise the needed CPU time. MwI was introduced recently [1], and it may reduce information entropy [2, 3] as efficiently as the combination of the Burrows-Wheeler transform [4–6] followed by Move-To-Front [7, 8] or Inversion Frequencies transforms [9]. MwI can, therefore, be used as a replacement for these transforms in the preprocessing stage of data compression algorithms [8, 10]. Additionally, MwI requires considerably less programming code compared to the aforementioned combinations. In contrast to the Burrows-Wheeler transform, MwI can also support streaming data compression, such as audio or sensory data. These data are processed frequently by embedded systems that have very limited computer resources. MwI is conceptually simple and its expected time complexity is linear in regard to the length of the sequence being transformed. Despite this, it requires careful implementation, especially in cases where it will run on modest computer hardware.

Three different C++ implementations of MwI are introduced in this chapter. For the purpose of experiments, they were also realised in Java. All six implementations are then run on three computer platforms: a personal computer with Windows and Linux, and a single-board computer Raspberry Pi running Linux. The implementations were tested on 32 benchmark greyscale raster images of different resolutions and contexts. The main contributions of the chapter are as follows:

- Highlighting the awareness of computer scientists and programmers that better computers do not solve the problems of shortening CPU time without careful implementation;
- Providing information about the CPU time spent by programming code written in two popular programming languages, C++ and Java;
- Testing the efficiency of implementations on three computer platforms.

The chapter is organised in six sections. Section 2 provides a brief overview of the previous works. Move with Interleaving, along with the Move-To-Front transform, is introduced briefly in Section 3. Three implementations of Move with Interleaving are described in Section 4. The results of the experiments are given in Section 5. Section 6 concludes the paper.

2. Previous works

In the past, careful implementation of algorithms was a strong focus, due to considerably limited computer resources. Scientists sought ways to achieve the results faster in practically all aspects of computer usage. Here are just a few examples:

Clark [11] proposed an optimisation of the CLEAN algorithm for setting radio astronomy aperture data. He analysed the algorithm and introduced two simplifications. Instead of processing the full image in every iteration, he suggested focusing on small regions containing the brightest features, a step referred to as the minor cycle. After several minor cycles, a major cycle updates the residual image globally, by applying corrections accumulated during the minor cycles. This approach offers a balanced trade-off between computational efficiency, memory requirements and accuracy, and is accepted widely in practice. Hadjiconstantinou and Christofides [12] proposed an efficient implementation for finding the K-shortest paths in undirected graphs with non-negative edge weights. They developed an advanced data structure for maintaining the long paths and limited the generation of non-promising paths. By experiments, the authors confirmed that the algorithm's CPU time grew linearly with respect to K, and quadratically with respect to the graph size. These optimisations, including better data structure and reducing the amount of data to be processed, make the algorithm suitable for applications requiring the determination of K-shortest paths, offering a good balance between computational cost and different graph configurations. Dorado et al. [13] applied the Analytic Hierarchy Process (AHP) to select the most suitable software for engineering education. While the chapter does not focus directly on computational optimisation, the authors introduced a systematic method for evaluating various software solutions based on multiple criteria. AHP optimises the selection process by reducing subjectivity and enhancing the consistency of the evaluation. Several optimisations have been proposed for the minimum-cost flow algorithm by Goldberg [14], including Scaling and refinement, Push-relabel optimisation, Price update and heuristics and Set-relabel operation. First, the method scales down the error threshold in each iteration, improving the flow gradually, bringing the solution closer to optimality. Push-relabel refines the order of the operations, enhancing performance in practical cases by reducing unnecessary steps. Price updates and heuristics represent the main optimisation, where the global prices of nodes are updated more efficiently and systematically, compared to the traditional local updates. Lastly, the Set-relabel operation updates the prices of a set of nodes simultaneously, instead of one at a time. These optimisations improve the computational efficiency of the algorithm, leading to a significant reduction in CPU time. Shi et al. [15] proposed two algorithms to improve the efficiency and accuracy of generating NURBS surfaces for 3D ship hull modelling. The first, the Uniform Continuity Generation algorithm, deals with handling straight-line segments efficiently, while the second, the Hybrid Surface Generation, optimises the surface control parameters to speed up the design process by avoiding undesirable artefacts. These algorithms offer a balance between computational efficiencies at the precision required for ship hull design. Lippert et al. [16] suggested a space-efficient implementation of the Burrows–Wheeler transform for genomic data. Their approach reduces memory usage for constructing the compressed suffix array, a critical data structure for sequence matching. By optimising the space complexity, they achieved a reduction in memory usage by a factor of two for the nucleotide alphabet, while maintaining the same asymptotic time complexity. These improvements enabled more efficient genomic data processing with reduced memory requirements. Žalik [17] developed an efficient Delaunay triangulation algorithm based on the sweep-line paradigm. The optimisation was achieved in two ways: First, reducing the number of legalisations of non-Delaunay triangles through heuristics, lowering the computational cost by nearly two-thirds. Second, by using a one-dimensional range search after projecting the considered point onto the advancing front, eliminating the need for two-dimensional searching data structures. The memory efficiency was reduced significantly because of this. These improvements resulted in a speed-up of up to 400% compared to previous methods. Lukač et al. [18] developed a high-resolution solar potential estimation method for large-scale urban areas using remote sensing data. They introduced an optimised anisotropic diffuse irradiance algorithm that accounts for environmental obstructions. Due to high computational demand, the algorithm was accelerated with GPGPU technology, achieving up to 70 times speed-up compared to multithreaded CPU implementation. This approach balances computational efficiency, accuracy, and the ability to handle the complexity of urban environments efficiently.

Today, however, there are often illusions that this problem is no longer relevant. The compilers generate efficient executable code and offer additional optimisation features. Devkota et al. [19] reported recently on optimisations of binary code, while a detailed description of the compiler design is provided in Ref. [20]. Better personal computers are, today, equipped with gigabytes of on-board memory, and their CPUs can perform calculations at a rate of hundreds of megaflops. Basford et al. [21] conducted a performance analysis of single-board computer clusters. However, with better computers, even more data are produced every day. The challenges of processing big data were, for example, considered by Khan et al. 10 years ago [22].

A huge amount of data require special (and expensive) computer installations. Stunkel et al. [23] wrote about the high-speed networks of the Summit and Sierra supercomputers. GPGPU installations are less expensive; however, they require a different approach to their efficient programming. The management of their workloads in virtualisation was presented by Iserte et al. [24]. The cloud-based architectures are another possibility for maintaining and processing today's data. Khriji [25] et al. designed and implemented a cloud-based event-driven architecture for real-time data processing in wireless sensor networks. Unfortunately, the majority of users do not have access to such installations; indeed, they operate modest personal computers, processing manageable amounts of data, for which they expect to receive results quickly. This is why even a short segment of programming code, executed frequently, should be understood, not only arithmetically, but also from an implementation perspective. Programming code could be optimised in various aspects, such as reducing spent CPU time, minimising the required memory, or lowering energy consumption. As modern processors contain more cores, parallelisation may reduce the processing time significantly. Žlaus and Mongus [26] developed a method for computing geodesic morphological operations on the CPU. Geodesic operations are essential in various applications, like image segmentation and surface analysis, and are in practice performed on raster images, which may have extreme resolutions. The authors achieved efficient computation of long chains of elementary filters using multicore and Single Instruction Multiple Data (SIMD) processing. In comparison with the related methods, up to 100 times faster computation of common geodesic operators is achieved in this way, allowing for real-time processing of up to 1500 filters long chains. Qiu et al. [27] introduced an optimisation method for Phase-Change Memory (PCM) in cloud computing environments using a genetic algorithm. The optimisation aims to improve both energy efficiency and system performance by allocating resources optimally. A genetic algorithm was applied to find a balance between energy consumption and system performance. Albers [28] explored several strategies for optimising energy use in computer systems, focusing particularly on energy management during idle periods. This is especially relevant in contexts such as cloud computing and mobile devices, where minimising energy consumption without sacrificing performance is crucial. Albers examined offline and online algorithms that aim to optimise energy use, with a focus on transitioning between active and sleep states efficiently. The key contribution is the development of competitive algorithms, which balances energy consumption and computing performances effectively.

2.1 Move-to-front transform

Move-To-Front (MTF) is one of the self-organising list strategies [7, 29–31] used in caching algorithms [32] and data compression [33, 34]. Let $X = \langle x_i \rangle, 0 \leq i < n - 1$, represent the input sequence of n elements x_i from an alphabet Σ_X, where Σ_X consists of m elements. MTF reads the elements $x_i \in X$ sequentially and maps them to the domain of natural numbers, including 0, to obtain the output sequence $Y = \langle y_i \rangle, 0 \leq i < n, y_i \in \Sigma_Y = \{0,1,2, \cdots, m - 1\}$. The list L is employed in this mapping process and it is populated with all the elements from Σ_X initially. The algorithm finds the position l, $0 \leq l < m - 1$, in L, where $x_i = L_l$ and sends l to Y. After that, the positions of all the elements L_e in L between $0 \leq e < l$ are increased, while x_i is placed in front of L. The pseudocode for the MTF transform is shown in Algorithm 1. The main loop contains only three functions. The first one, on Line 7, iterates through the elements in L and stops when $L_l = x_i$. It is expected that x_i is found in the constant time $O(1)$, since MTF is a self-organising list. The function in Line 8 terminates in $O(1)$, as it just writes l to an array representing Y. The most time-consuming task is performed by the last function in Line 9, which needs to replace the first $l - 1$ values in L. It can be realised in various ways. Four different implementations were proposed and analysed in Ref. [35].

MTF is used most frequently in data compression [30, 31, 33], as it reduces the information entropy of X if X exhibits local correlations. For example, a sequence of the same elements is converted into a sequence of zeros; a sequence of alternating elements gives a sequence of ones; a sequence of alternating triples results in a sequence of twos and so on. This is the reason why MTF is used typically after BWT, which tends to group the same elements close together [5, 6].

Algorithm 1. Pseudocode of MTF transform.

```
1.  function MTF(X, n, Σ_X, m)            ▷ returns the transformed sequence Y
2.                                        ▷ X: input sequence with length n
3.                                        ▷ Σ_X: alphabet with length m
4.      i = 0
5.      L = Initialise(Σ_X, m)            ▷ insert all elements from Σ_X to L
6.      for i ← 0 to n do
7.          l = GetPosition(L, x_i)
8.          Y = AddToY(Y, l)
9.          L = MoveUp(L, l)   ▷ increase positions of elements in L between 0 and l
10.         L_0 = x_i                     ▷ move x_i to the front
11.     end for
12.     return Y                          ▷ return the transformed sequence
13. end function
```

2.2 Move with interleaving

The *Burrows-Wheeler transform, (BWT)* should be used before MTF to enhance the local correlations of X, as mentioned in the previous subsection. BWT was introduced in 1994 [4] and is a well-studied transform [8]. Unfortunately, it has $O(n^2\log(n))$ time complexity, which is why X should be split into small blocks to achieve a fast enough response [4]. A direct correlation between suffix array and the BWT was determined later [5]. Straightforward implementation of the suffix array also works, unfortunately, in $O(n^2\log(n))$ [36]. Fortunately, more algorithms were developed for constructing a suffix array of X in $O(n)$ [37–39]. This also enables obtaining the BWT in linear time. However, the implementation of these algorithms is relatively demanding. This is why it would be desirable to replace the pipeline of the three algorithms—an algorithm for constructing the suffix array, an algorithm to generate BWT from the suffix array, and the MTF algorithm—with only one compact algorithm. It was shown recently that, at least for the domain of Raster Images, this can be done by the transform named *Move with Interleaving (MwI)* [1]. Pixels from a continuous-tone image are arranged into a sequence X at first. Pixels in the close neighbourhood of the considered pixel are, in the majority of cases, similar, but not completely the same. However, this similarity is sufficient for MTF to produce small indices y_i. But, when a region of different pixel values is encountered, MTF has to generate many large indices y_i before bringing enough values from this region close to the front of L. This has, unfortunately, a negative effect on the reduction of the information entropy, and this is why BWT should be applied before MTF. Because of this, MwI monitors the value $y_i = MTF(x_i)$ and reacts when y_i exceeds the given threshold t. MwI assumes that an area with different values is found and that the values similar to x_i will follow. Because of this, it inserts, in addition to x_i, $2t$ interleaved values around x_i in front of L, that is the values from a generated auxiliary sequence $A = \langle x_i, x_i + 1, x_i - 1, x_i + 2, x_i - 2, \cdots, x_i + t, x_i - t \rangle$.

The MwI pseudocode is shown in Algorithm 2. The algorithm first populates the list L in Line 5. It uses the first element from X and interleaves it with $2t$ values. These values are inserted in front of L, and the remaining values from Σ_X are filled in alphabetical order. The first element x_0 is stored immediately in the output Y in Line 6, as the decoder should perform the same initialisation of L. The algorithm then enters the main loop and processes $X = \langle x_i \rangle, 1 \leq i < n$, sequentially. The position l of x_i is determined in L (Line 8). It checks after that whether l is within the given threshold t. If so, the MTF procedure is performed in Lines 11 and 12; otherwise, the algorithm fills the auxiliary sequence A with interleaved elements around x_i in Line 14. The elements from A are then inserted in L in the MTF-style within Lines 16 and 17.

Algorithm 2. Pseudocode of MwI transform.

```
1. function MwI(X, n, Σ, m, t)                ▷ returns the transformed sequence Y
2.                                             ▷ X: input sequence with length n
3.                                             ▷ Σ: alphabet with length m
4.                                             ▷ t: threshold
5.   L = InitialiseL(X, n, Σ, m, t)            ▷ initialisation of L according to x_{i=0}
6.   Y_0 = x_0                                 ▷ store x_{i=0}
7.   for i ← 1 to n do                         ▷ for all other x_i
8.     l = GetPosition(L, x_i)                 ▷ find the position of x_i in L
9.     Y = AddToY(Y, l)                        ▷ store the position in Y
10.    if l ≤ t then                           ▷ if the position is smaller than t
11.      L = MoveUp(L, l)                      ▷ perform MTF
12.      L_0 = x_i
13.    else                                    ▷ otherwise perform MwI
14.      A = FillA(x_i, t)                     ▷ fill temporary sequence T
15.      for k ← (2 × t) downto 0 do           ▷ execute MTF 1 + 2 × t times
16.        L = MoveUp(L, A_k)
17.        L_0 = A_k
18.      end for
19.    end if
20.  end for
21.  return Y                                  ▷ returns the transformed sequence
22. end function
```

Let us consider an example, where the first 10 elements of a hypothetical raster greyscale image with $\Sigma_X = \{0,1,2,\cdots,255\}$ are collected in sequence $X =$

$\langle 2,4,1,1,3,151,150,153,152,152, \cdots \rangle$. The smaller values correspond to dark, while the larger ones to bright pixels. In our case, the first five elements are dark, while the remaining five belong to a brighter region. However, in both regions, the values are not the same; they are just similar, which is a characteristic of photographic images. The process of MTF transform is shown in **Table 1**. It can be seen that MTF requires many steps to adapt to the new situation, that is to pixels from the bright region. This is why the resulting sequence $Y = \langle 2,4,3,0,4,151,151,153,153,0 \rangle$ still contains many large values. However, MTF still managed to reduce the information entropy from $H(X) = 3.122$ to $H(Y) = 2.522$.

Table 2 shows the process of MwI transform when $t = 3$. It can be seen that MwI adapts the content of L efficiently to the new situation. The obtained transform $Y = \langle 0,3,3,0,3,151,2,3,3,0 \rangle$ has information entropy of $H = 1.685$ and contains considerably smaller values of its elements, which is beneficial in further compression steps.

Finally, let us analyse the time complexity of MwI. $l > t$ in each iteration for the worst-case scenario. Therefore, for each $x_i, 1 \leq i < n,$ the algorithm fills A in time $T_A = 2t + 1$ and then performs the MTF $2t + 1$ times. MTF has the expected time complexity $O(n)$, and, as a result, the expected execution time of MwI is $T_{MwI} = (2t + 1)O(n)$. Since $t \leq m < < n$, it can be concluded that MwI's time complexity is the same as MTF, i.e. $O(n)$.

3. Presentation of implementations

Three different implementations of MwI, denoted as *I-1*, *I-2*, and *I-3*, are considered in this section. The programming code is presented in C++ [40], with a note that the paper does not make use of its object-oriented features.

3.1 Implementation I-1

List L is represented as the standard C++ vector in this implementation. Algorithm 3 shows the function that initialises L. This initialisation assumes that the alphabet represents the numbers from the range $[lBound, uBound]$. The last parameter N in function InitL_I controls the number of elements to be inserted. It is set to $uBound - lBound + 1$ initially. Variables u and l contain the incremental interleaving values around FirstEl.

The main function of implementation *I-1* is shown in Algorithm 4. The position l of each x_i in L is determined in Line 10 and then stored in Line 11. After that, l is tested (Line 12): if it is smaller than t, the MTF is executed in Lines 13–15; otherwise, more elements interleaved around x_i should be moved-to-front of L. The auxiliary vector A is filled with at most 1 + 2 t interleaved values by calling the function InitL_I in Line 18. In Algorithm 3, this time the parameter N gets the value 2 t. After that, the content of L is updated in Line 19.

Finally, let us explain the updating procedure for L, when more elements should be moved to the front (see Algorithm 5). In Line 4, 1 + 2 t interleaved values from vector A are first copied to a temporary vector Temp. The remaining values of L are added incrementally within the Lines 6–17 to Temp if they have not yet been added in Line 4. When the for-loop terminates, vector Temp contains the updated status of L, which is then returned in Line 18 for use in the next iteration of Algorithm 4.

3.2 Implementation I-2

In implementation *I-2*, the data structure used to store L is once again a standard vector. Consequently, the initialisation process remains identical to that given in

X		**/**	**2**	**4**	**1**	**1**	**3**	**151**	**150**	**153**	**152**	**152**
L	0	0	2	4	1	1	3	151	150	153	152	152
	1	1	0	2	4	4	1	3	151	150	153	149
	2	2	1	0	2	2	4	1	3	151	150	153
	3	3	3	1	0	0	2	4	1	3	151	150
	4	4	4	3	3	3	0	2	4	1	3	151
	5	5	5	5	5	5	5	0	2	4	1	3
	6	6	6	6	6	6	6	5	0	2	4	1
	:	:	:	:	:	:	:	:	:	:	:	:
	149	149	149	149	149	149	149	148	147	146	145	145
	150	150	150	150	150	150	150	149	148	147	146	146
	151	151	151	151	151	151	151	150	149	148	147	147
	152	152	152	152	152	152	152	152	152	149	148	148
	153	153	153	153	153	153	153	153	153	152	149	149
	:	:	:	:	:	:	:	:	:	:	:	
	255	255	255	255	255	255	255	255	255	255	255	255
Y		/	2	4	3	0	4	151	151	153	153	0

Table 1.
An example of MTF, where the first column shows the initial status of L.

X		**/**	**2**	**4**	**1**	**1**	**3**	**151**	**150**	**153**	**152**	**152**
L	0	2	2	4	1	1	3	151	150	153	152	152
	1	3	3	2	4	4	1	152	151	150	153	153
	2	1	1	3	2	2	4	150	152	151	150	150
	3	4	4	1	3	3	2	153	153	152	151	151
	4	0	0	0	0	0	0	149	149	149	149	149
	5	5	5	5	5	5	5	154	154	154	154	154
	6	6	6	6	6	6	6	148	148	148	148	148
	:	:	:	:	:	:	:	:	:	:	:	:
	149	149	149	149	149	149	149	142	142	142	142	142
	150	150	150	150	150	150	150	143	143	143	143	143
	151	151	151	151	151	151	151	144	144	144	144	144
	152	152	152	152	152	152	152	145	145	145	145	145
	153	153	153	153	153	153	153	146	146	146	146	146
	:	:	:	:	:	:	:	:	:	:	:	
	255	255	255	255	255	255	255	255	255	255	255	255
Y		/	0	3	3	0	3	151	2	3	3	0

Table 2.
MwI transform: an example.

Algorithm 3. The primary drawback of implementation *I-1* is in the update function outlined in Algorithm 5, which contains two nested loops. The first loop iterates through all the elements of L. The nested *while* loop scans the auxiliary vector A to verify whether the element L[i] also exists in A. If not, the element is added to the temporary vector Temp. Although the UpdateL function is not called for each element being transformed, and the length of the vector A is expected to be small, $1 + 2t < < m$, it opens up room for an improvement. Namely, the nested **while** loop can be omitted by introducing a lookup table LUT. Its purpose is similar to the lookup table in the Counting-sort algorithm [41]. In our case, LUT consists of m flags set to FALSE initially. The flags in LUT are set during the creation of the auxiliary array A, that is each flag is raised at the position LUT_{A_i}. During the updating of L, for each element in L it checks the status of the flag at the location LUT_{L_i}. The element is added into the temporary vector if the flag is FALSE. Details, including the construction of vector A and the updating of L, are presented in Algorithm 6. An array of flags LUT is allocated in Line 6. It consists of m elements, that is the size of the alphabet Σ_X. In this implementation, it is assumed that the alphabet Σ contains elements from the interval $[0, m - 1]$. If this is not the case, a simple mapping of the alphabet's element into this interval should be done before calling the function MwIwithLUT. When the position l of the considered element $x_i > t$, all flags in LUT are set to FALSE in Line 20. The flag belonging to l is set to TRUE after that in Line 21. The same is applied for all the remaining interleaved values in Lines 27 and 31. The update of L is started in Line 35 after A has been prepared. The **for** loop operates for all elements of Σ_X. For each element of L, it is checked whether its flag in LUT is set (Line 36). If it is not, the considered element is inserted into A. Finally, A is copied into the new L in Line 38.

Algorithm 3. Implementation *I-1*: initialization of *L*.

```
vector<int> InitL_I(int lBound, int uBound, int FirstEl, int N)
        ▷ lBound, uBound: the values of the first and the last element from Σ
                                    ▷ FirstEl: the value of the first element x0
                            ▷ N: themaximal number of elements to be inserted
{
  vector<int> L(N+1);
  L[0] = FirstEl;                   ▷ FirstEl becomes the first element of L
  int u = FirstEl + 1;                        ▷ incremental interleaving value
  int l = FirstEl – 1;                        ▷ decremental interleaving value
  int counter = 0;
  int pos = 1;
  while (counter++ < (int)(N/2)) {
    if (u <= uBound)
      L[pos++] = u++;
    if (l >= lBound)
      L[pos++] = l– –;
  }
  return L;
}
```

Algorithm 4. Implementation *I-1*: MwI main function.

```
vector<int> MwI_I(vector<int> X, int t, int lBound, int uBound,
int vector<int>& L)
{
 int i, j, l;
 vector<int> A;
 vector<int> Y(X.size());                          ▷ allocate space for Y
 Y[0] = X[0];                                      ▷ first element in Y is X[0]
 for (i = 1; i < (int)X.size(); i++){   ▷ for all elements of X except the first
  l = 0;
  while (X[i] != L[l]) l++;                        ▷ find the position l of x_i in L
  Y[i] = l;                                        ▷ store position l in the result
  if (l <= t) {
   for (j = l – 1; j >= 0; j– –)                   ▷ move-to-front
   L[j + 1] = L[j];
   L[0] = X[i];
  }
  else {                                           ▷ move-to-front more elements
   A = InitL_I(lBound, uBound, X[i], 2 * t);       ▷ fill the auxiliary array
   L = UpdateL(A, L);                              ▷ update L with elements from A
  }
 }
 return Y;
}
```

Algorithm 5. Implementation *I-1*: *L* update function.

```
vector<int> UpdateL(vector<int> A, vector<int> L)
{
 int NotExistInA, i, j;
 vector<int> Temp = A;               ▷ first Temps' elements are the same as A
 Temp.reserve(L.size());               ▷ allocate space for all elements from L
 for (i = 0; i < (int)L.size(); i++) {     ▷ for all elements of the alphabet
  NotExistInA = 1;
  j = 0;
  while (NotExistInA && (j < (int)A.size())) {
   if (L[i] = = A[j])                ▷ check if an element from L exists in A
   NotExistInA = 0;                                  ▷ mark that it exists
   else
   j++;                          ▷ otherwise check the next element from A
  }
```

```
      if (NotExistInA)
      Temp.push_back(L[i]);                ▷ if the element does not exist, add it
    }
    return Temp;                           ▷ Temp becomes the new L
}
```

3.3 Implementation I-3

Implementation *I-3* utilises a dynamically linked list of records to represent *L*. Each record has only two components: el, containing the value, and a pointer next that points to the subsequent record. Concerning MTF, it seems that this data structure could offer benefits, as there is no requirement to move the elements at positions 0 through $l - 1$.

Let us consider an example of *L* shown in **Figure 1a**, containing 10 elements. Suppose the element that should be moved to the front of *L* has the value 7. The record containing this value has already been found and is pointed to by pointer w; its preceding pointer is denoted as wPrev. The new status of *L* is established in constant time by repointing just three pointers, as illustrated in **Figure 1b**. The details are provided in Algorithm 7.

Algorithm 6. Implementation *I-2*: MwI with LUT.

```
vector<int> MwIwithLUT(vector<int>& X, int t, int lBound, int uBound,
int vector<int>& L)
{
  vector <int> Y(X.size());                      ▷ allocate space for Y
  int LUTSize = (int)L.size();
  vector<int> LUT(LUTSize);                      ▷ allocate space for LUT
  Y[0] = X[0];                                   ▷ the first element in Y is X[0]
  for (int i = 1; i < (int)X.size(); i++){ ▷ for all elements of X ecept the first
    int l = 0;
    while (X[i] != L[l]) l++;                    ▷ find the position l of x_i in L
    Y[i] = l;                                    ▷ store position l in the result
    if (l <= t) {
      for (j = l – 1; j >= 0; j– –)              ▷ move-to-front
      L[j + 1] = L[j];
      L[0] = X[i];
    }
    else {  ▷ construct interleaved elements around L[l] and store them into A
      vector<int> A(L.size());
      A[0] = L[l];                               ▷ the first element in A is L[l]
      fill(LUT.begin(), LUT.end(), 0);           ▷ set all flags in LUT to 0
      LUT[l] = 1;                                ▷ set the flag in LUT
      int uv = L[l] + 1;
      int lv = L[l] – 1;
```

```
            int c = 1;
            for (int j = 0; j < t; j++) {                ▷ insert at most 2t elements in A
              if (uv <= uBound) {
                LUT[uv] = 1;                                        ▷ set the flag in LUT
                A[c++] = uv++;
              }
              if (lv >= lBound) {
                LUT[lv] = 1;                                        ▷ set the flag in LUT
                A[c++] = lv– –;
              }
            }
            for (int k = 0; k < LUTSize; k++)     ▷ for all elements of the alphabet
            if (!LUT[L[k]])              ▷ If the flag in LUT is not set, add an element
            A[c++] = L[k];
            L = A;                                            ▷ A is copied to the new L
          }
      }
      return Y;
  }
```

Algorithm 7. Implementation *I-3*: MTF.

```
ListType* UpdateMTF(ListType* L, int x, int& l)
{       ▷ x should be moved to the front of L; function returns position l and
        updated L
  if (L->el == x) {                                         ▷ x is already on the top
     l = 0;                                                          ▷ set its position
     return L;                                                 ▷ return the existing L
  }
  ListType *w, *wPrev;
  w = wPrev = L;
  l = 0;
  while (w–>el != x) {                             ▷ work until x is not found in L
    l++;                                                   ▷ increment the position
    wPrev = w;                                       ▷ move pointers for one record
    w = w–>next;
  }
  wPrev–>next = w–>next;                                 ▷ relink the pointers in L
  w–>next = L;
  L = w;
  return L;                                                            ▷ return new L
}
```

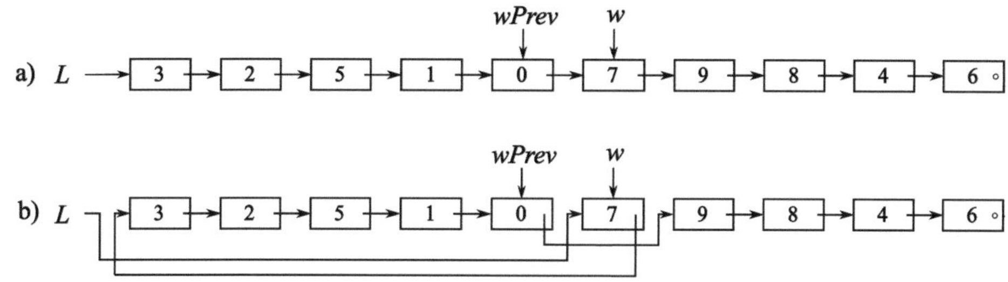

Figure 1.
L reconfiguration when a dynamic linked list is used.

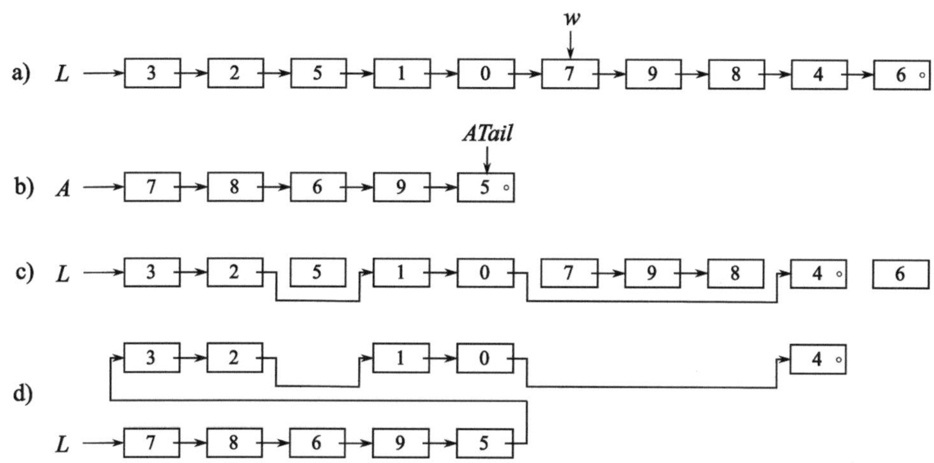

Figure 2.
L reconfiguration at MwI when a dynamic linked list is used.

The concept of how the MwI transformation can be realised using a dynamic linked list is shown in **Figure 2**. Let us suppose that $x_i = 7$ is again the considered value in L, which is shown in 2a. In addition, let us suppose that $t = 2$. The position of $x_i = 7$ is at $l = 5$, and as $l > t$, the auxiliary list A of 1 + 2 t values should be appended in front of L. The auxiliary list A is shown in **Figure 2b**. L is inspected, and all its records holding values within the range $[x_i - t, x_i + t]$ are omitted (see **Figure 2c**). Finally, L and A are linked together using the pointer $ATail$; the result is shown in **Figure 2d**. We can observe that 1 + 2 t new records have been created, and an equal number of them have been deleted. Memory allocation and release are, unfortunately, computationally expensive operations and should be avoided whenever possible [42]. Therefore, records that should be released are placed in a stack named *Pool* in the continuation. Whenever a new record is needed during the process of constructing A, it is taken from the *Pool*. Details of the implementation can be found in Algorithm 8. In Line 9, the function PreparePool inserts 1 + 2 t dynamically allocated records into the stack pointed to by Pool. Function GetPosition (Line 11) performs the same tasks as the programming code in Lines 8 to 14 of Algorithm 7. After that, the records containing the values from the interval $[x_i - t, x_i + t]$ are inserted into Pool by the function in Line 15. The function ConstructA (Line 16) constructs list A with interleaved values around the value x_i using the records from the Pool. It returns pointers A and $ATail$. In Lines 17 and 18, lists L and A are merged into a new list L. In Lines 21, 22, and 23, the record is moved in front as explained earlier. All implementations are accessible at Ref. [43].

4. Experiments

All three proposed implementations of MwI were tested on 32 benchmark 8-bit greyscale images. The images can be seen in Ref. [1], and, in this chapter, they are denoted in the same way. The values of pixels were arranged into the sequence X in the scan-line order. All three implementations were coded in both C++ and Java, as the most popular programming languages, and they were executed on three different platforms:

- *P1*: A personal computer with AMD Ryzen 97950X3D 16-Core Processor clocked at 4.20 GHz, equipped with 64 GB of RAM, running the Windows 11 operating system. The code was compiled using MS Visual Studio, Version 17.4.2. OpenJDK-8-JRE was used for Java.

- *P2*: The same personal computer as in P1, but using Linux Kernel 6.6.8. with the Manjaro distribution. The code was compiled with G++ compiler Version 13.2.1. for C++. OpenJDK-8-JRE was used for Java again.

- *P3*: The Raspberry Pi 2 Model B with ARM Cortex-A7 CPU and 1 GB RAM running Raspbian GNU/Linux 11 (bullseye). The C++ code was compiled with gcc version 10.2.120210110 (Raspbian 10.2.1–6 + rpi1), while OpenJDK-8-JRE was used for Java.

Table 3 illustrates the CPU time spent for all three implementations across the three platforms for code written in C++. All the tests were executed 10 times, and the deviations in results were within 1%. The results were indeed stunning. Implementation *I-2* was 562% faster than *I-1*. However, implementation *I-3* surpassed *I-2* by 334% and outperformed *I-1* by an astonishing 1883% on platform P1. Similarly, on platform P2, *I-3* was 1908% faster than *I-1* and surpassed *I-2* by 246%. Interestingly, despite using the same computer, platform P2 outperformed platform P1 consistently. While *I-1* and *I-3* differed by 4% and 5%, respectively, platform P2 surpassed P1 for *I-2* significantly by 42%. The pattern repeated on platform P3, but with a more pronounced effect. *I-3* was, this time, faster than *I-1* by more than 2300%, and by more than 550% from *I-2*. However, in this case, the absolute spent CPU time also becomes important. The user should wait for 7 and a half minutes to transform all 32 images using implementation *I-1*, while *I-3* requires only 19.5 seconds.

X	n	P1			P2			P3		
		I-1	*I-2*	*I-3*	*I-1*	*I-2*	*I-3*	*I-1*	*I-2*	*I-3*
(1)	262,144	0.454	0.081	0.027	0.435	0.058	0.027	7.382	1.753	0.373
(2)	414,720	0.177	0.033	0.013	0.171	0.025	0.013	2.907	0.726	0.177
(3)	414,720	0.528	0.095	0.036	0.505	0.068	0.033	8.735	2.160	0.447
(4)	262,144	0.392	0.068	0.030	0.370	0.048	0.027	6.420	1.514	0.356
(5)	262,144	0.504	0.091	0.038	0.479	0.063	0.035	8.262	2.053	0.467
(6)	414,720	0.196	0.037	0.015	0.189	0.027	0.014	3.210	0.816	0.200
(7)	414,720	0.289	0.055	0.019	0.276	0.038	0.018	4.744	1.240	0.265
(8)	65,536	0.065	0.012	0.005	0.060	0.008	0.004	1.041	0.256	0.057
(9)	262,144	0.235	0.046	0.017	0.221	0.031	0.016	3.751	0.888	0.217
(10)	245,760	0.162	0.031	0.011	0.153	0.020	0.009	2.614	0.638	0.139
(11)	181,000	0.205	0.036	0.014	0.195	0.024	0.012	3.328	0.764	0.162
(12)	245,760	0.177	0.034	0.011	0.167	0.022	0.009	2.861	0.711	0.142
(13)	414,720	0.339	0.064	0.020	0.321	0.045	0.019	5.551	1.407	0.274
(14)	414,720	0.387	0.070	0.021	0.368	0.049	0.020	6.355	1.576	0.289
(15)	262,144	0.214	0.040	0.014	0.202	0.027	0.013	3.469	0.881	0.189
(16)	1,745,280	1.970	0.342	0.105	1.880	0.240	0.100	32.338	7.732	1.387
(17)	5,474,148	1.801	0.337	0.114	1.715	0.245	0.107	29.313	7.425	1.534
(18)	1,048,576	1.210	0.211	0.066	1.150	0.144	0.060	19.654	4.615	0.805
(19)	995,520	0.474	0.086	0.030	0.454	0.060	0.026	7.717	1.830	0.356
(20)	387,228	0.268	0.047	0.014	0.256	0.032	0.011	4.371	1.022	0.177
(21)	393,216	0.261	0.048	0.019	0.249	0.034	0.016	4.260	1.057	0.222
(22)	262,144	0.010	0.004	0.001	0.008	0.001	0.001	0.162	0.059	0.027
(23)	154,401	0.200	0.037	0.016	0.192	0.024	0.014	3.263	0.771	0.191
(24)	393,216	0.235	0.041	0.015	0.225	0.029	0.013	3.826	0.921	0.181
(25)	245,760	0.228	0.041	0.015	0.216	0.028	0.013	3.721	0.910	0.188
(26)	262,144	0.238	0.044	0.017	0.227	0.030	0.014	3.865	0.952	0.195
(27)	2,073,600	1.959	0.351	0.138	1.872	0.236	0.161	31.632	7.489	1.707
(28)	393,216	0.273	0.049	0.017	0.264	0.033	0.015	4.445	1.070	0.208
(29)	4,271,400	1.507	0.286	0.091	1.450	0.204	0.084	24.726	6.070	1.204
(30)	17,448,000	12.131	2.091	0.481	11.802	1.477	0.457	201.470	47.227	6.975
(31)	245,760	0.188	0.035	0.015	0.180	0.024	0.013	3.052	0.750	0.169
(32)	414,720	0.266	0.053	0.017	0.252	0.035	0.015	4.339	1.137	0.216
Sum		27.543	4.896	1.462	26.504	3.429	1.389	452.784	108.420	19.496

Table 3.
CPU time spent [s] when the C++ programming language was used.

Algorithm 8. Implementation I-3: MwI.

```
vector<int> MwIwithList(vector<int>& X, int t, int lBound, int uBound,
ListType* L)
{           ▷ parameters X, t, lBound, uBound are the same as in Algorithm 4
        ▷ L is the pointer to elements of Σ_X with interleaved values in regards
          to X[0]
  ListType *A, *ATail, *w, *wPrev, *Pool;
  int i, l;
  vector<int> Y(X.size());                              ▷ allocate space for Y
  Y[0] = X[0];                                ▷ the first element in Y is X[0]
  Pool = PreparePool(t);                              ▷ allocate 1+2t records
  for (i = 1; i < (int)X.size(); i++) {   ▷ for all elements of X except the first
    l = GetPosition(X[i], &wPrev, &w);  ▷ return l and get *w and *wPrev
    Y[i] = l;                                                      ▷ store l
    if (l > 0) {                                     ▷ if l is 0, nothing to do
      if (l > t) {
      L = RemoveFromL(X[i], t, L, Pool)                  ▷ Sent recors to Pool
      ConstructA(X[i], &Pool, &A, &ATail);  ▷ A uses records from Pool
      ATail->next = L;                                        ▷ Link L and A
      L = A;                                    ▷ set L on the top of the list
    }
    else {                                     ▷ MTF part of rearranging L

      wPrev->next = w->next;
      w->next = L;
      L = w;
    }
  }
 }
 return Y;                                                    ▷ return new L
}
```

The results for the Java implementation are provided in **Table 4**. Implementation *I-3* again surpassed *I-1* and *I-2* by more than 1400% and 680%, respectively, on platform P1. Similarly, on platform P2, *I-3* outperformed *I-1* by over 1270% and *I-2* by 580%. Platform P2 was faster than platform P1 for *I-1* and *I-2*, but not for *I-3*. On platform P3, implementation *I-3* outperformed *I-1* by more than 6700%, and *I-2* by almost 1000%.

The C++ code was optimised by the -O3 compiler flag, which is known for achieving the highest levels of optimisation. The flags -XX:+OptimizeString-Concat and -XX: +UseCompressedOops were tried in Java. However, they did not have a noticeable effect. On all platforms the flag for controlling the Java heap size was set to 512 MB

X	n	P1			P2			P3		
		I-1	*I-2*	*I-3*	*I-1*	*I-2*	*I-3*	*I-1*	*I-2*	*I-3*
(1)	262,144	0.659	0.739	0.071	0.696	0.358	0.079	103.426	15.059	1.397
(2)	414,720	0.242	0.284	0.029	0.259	0.138	0.038	39.905	6.434	1.001
(3)	414,720	0.672	0.791	0.062	0.687	0.333	0.061	115.944	17.137	1.500
(4)	262,144	0.506	0.526	0.054	0.497	0.230	0.051	85.874	12.795	1.173
(5)	262,144	0.678	0.327	0.058	0.648	0.304	0.062	110.942	16.434	1.768
(6)	414,720	0.255	0.114	0.029	0.259	0.120	0.025	43.032	6.766	0.809
(7)	414,720	0.377	0.170	0.038	0.382	0.178	0.034	62.491	9.475	1.044
(8)	65,536	0.087	0.032	0.010	0.083	0.037	0.008	14.074	2.111	0.222
(9)	262,144	0.286	0.116	0.014	0.303	0.141	0.029	50.448	7.548	0.792
(10)	245,760	0.212	0.100	0.028	0.210	0.094	0.019	35.155	5.276	0.609
(11)	181,000	0.262	0.149	0.020	0.267	0.121	0.022	45.329	6.583	0.598
(12)	245,760	0.228	0.101	0.016	0.227	0.102	0.018	38.372	5.648	0.661
(13)	414,720	0.436	0.187	0.033	0.450	0.205	0.035	74.056	10.972	1.062
(14)	414,720	0.491	0.214	0.038	0.504	0.229	0.036	85.701	12.234	1.266
(15)	262,144	0.277	0.116	0.025	0.276	0.126	0.024	47.112	7.173	0.697
(16)	1,745,280	2.592	1.184	0.184	2.532	1.145	0.177	439.323	62.674	5.644
(17)	5,474,148	2.691	1.342	0.377	2.488	1.260	0.288	402.185	67.144	13.515
(18)	1,048,576	1.677	0.651	0.107	1.558	0.716	0.108	268.647	38.267	3.223
(19)	995,520	0.681	0.269	0.043	0.724	0.291	0.051	104.993	15.771	1.689
(20)	387,228	0.339	0.135	0.019	0.342	0.154	0.023	59.133	8.521	0.802
(21)	393,216	0.329	0.144	0.025	0.343	0.156	0.030	56.893	8.434	0.858
(22)	262,144	0.014	0.002	0.001	0.016	0.009	0.004	2.109	0.470	0.243
(23)	154,401	0.275	0.106	0.018	0.262	0.123	0.025	43.936	6.368	0.624
(24)	393,216	0.299	0.130	0.021	0.304	0.138	0.024	51.706	7.614	0.742
(25)	245,760	0.294	0.131	0.014	0.296	0.130	0.024	49.857	7.145	0.649
(26)	262,144	0.356	0.172	0.015	0.307	0.141	0.025	52.295	7.626	0.701
(27)	2,073,600	2.513	1.313	0.188	2.528	1.152	0.219	431.911	61.197	6.356
(28)	393,216	0.350	0.150	0.021	0.360	0.162	0.028	60.802	9.059	0.827
(29)	4,271,400	2.028	0.851	0.143	1.982	0.937	0.166	333.102	49.833	6.114
(30)	17,448,000	16.950	7.056	0.904	15.700	7.061	1.050	2733.081	386.570	33.050
(31)	245,760	0.241	0.127	0.019	0.247	0.113	0.024	41.698	6.001	0.624
(32)	414,720	0.333	0.381	0.020	0.345	0.164	0.029	58.023	8.572	0.872
Sum		37.630	18.110	2.653	36.082	16.568	2.836	6141.555	892.911	91.132

Table 4.
CPU time spent [s] when the Java programming language was used.

(−Xmx512m). The Raspberry Pi 2 Model B has 1 GB of RAM, and, by default, it allocates 250 MB for the heap.

As expected, the Java implementation was slower than C++ implementation (compare the results in **Tables 3** and **4**), the reason for this is discussed briefly in next section. Let us consider the results on the personal computer, that is on platforms P1 and P2. The most evident difference is at *I-2*, where platforms P1 and P2 lag behind the C++ implementation by more than 360% and 480%,

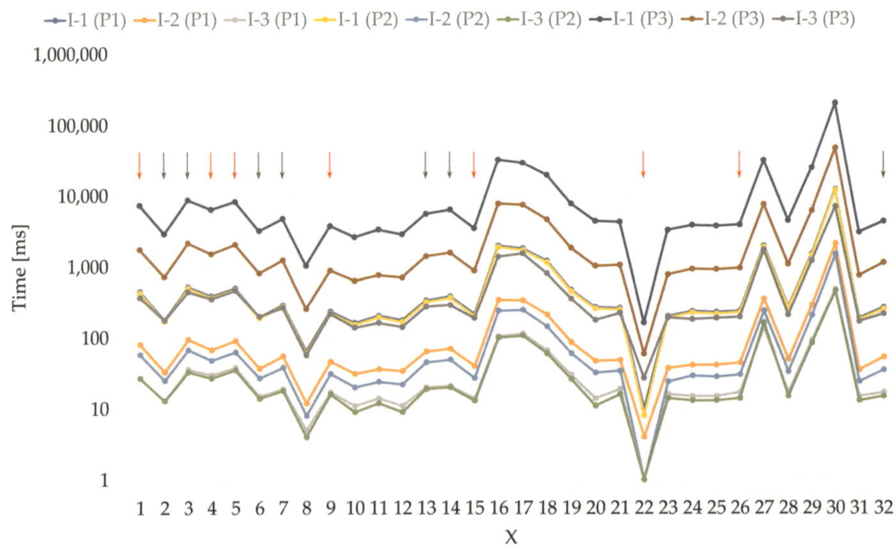

Figure 3.
CPU time spent when C++ was used for the implementation of MwI transform.

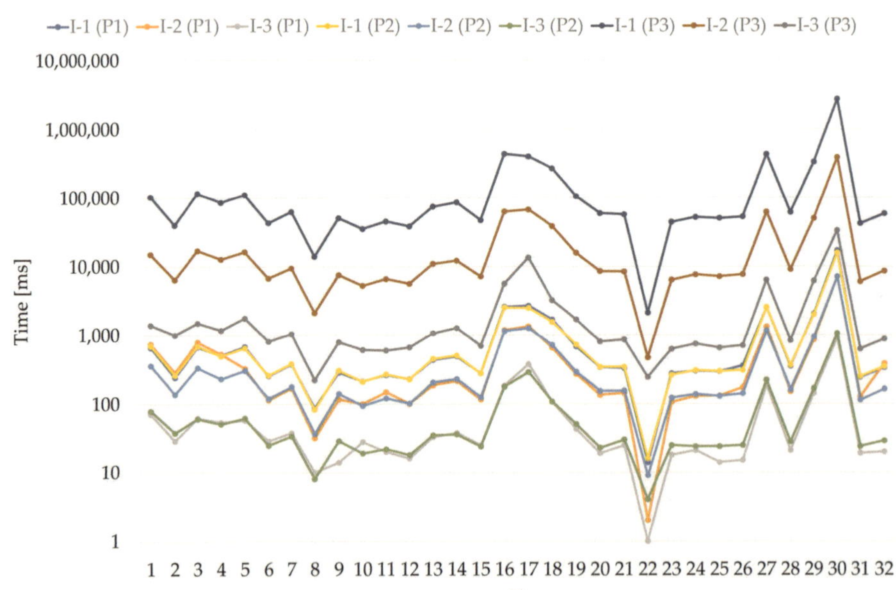

Figure 4.
CPU time spent by Java on executing MwI transform.

Figure 5.
Images (5) and (22) from the testing datasets.

respectively. Platform P2 was faster for *I-1* and *I-2*, while it was slightly slower for *I-3* (by less than 7%). The Java implementations were significantly slower than the C++ implementations on platform P3. Implementation *I-1*, written in Java, was slower by a hardly believable 1358%, *I-2* by more than 800%, and *I-3* by more than 460%. In case of *I-1* coded in Java, the user would need to wait for 1 hour and 42 minutes to transform all 32 images, while, for *I-3*, this process is completed in one and a half minutes.

Although MwI transform is simple to understand and its expected time complexity is $O(n)$, where n is the length of sequence X to be transformed, its actual CPU time depends significantly on various factors, including the implementation, programming language, and the execution platform. The graphs in **Figures 3** and **4** summarise the results of **Tables 3** and **4** for implementations in C++ and Java, respectively. Please notice the logarithmic scale of CPU time spent.

It can be seen that the relations between implementations do not change among the datasets and platforms. Implementation *I-3* clearly outperforms on all platforms implementations *I-1* and *I-2*, regardless of the programming language used. However, CPU time spent also depends on another factor. For example, datasets (1), (4), (5), (9), (15), (22), and (26), denoted by red arrows in the graph in **Figure 3**, have the same number of elements, that is 262,144, yet their running times differ considerably. Datasets (5) and (22) represent two extremes. Running implementation *I-1* in Java programming language on platform P1, processing dataset (5) requires 678 ms, while the processing of dataset (22) needs only 14 ms, which is almost 50 times faster. These two datasets are shown in **Figure 5**. The image on the left (dataset (5)) is characterised by very low local correlations between its pixels. On the contrary, the image on the right, representing dataset (22), exhibits extremely high local correlations. In dataset (5), $l > t$ appears to be true often, leading to frequent construction of the auxiliary array A with interleaved values and updating the list L with them. However, in dataset (22), $l \leq t$ is predominant, as the regions containing highly correlated values (i.e. pixels with the same colours) are large.

Datasets (2), (3), (6), (7), (13), (14), and (32), which also have the same number of elements (414,720), are marked with green arrows in the graph in **Figure 3**.

Deviations in spent CPU time are noticeable in this case as well. However, they are less evident, since none of these images are created artificially, and do not have huge regions of the same pixels (the images can be seen in [1]).

The Java Virtual Machine (JVM) ensures the execution of Java bytecode by a Just-In-Time compiler, thereby achieving platform independence [44]. However, this results in a slight decrease in efficiency on platforms P1 and P2. The algorithms considered in this study were implemented in the same way, with only minor language-specific differences. The automatic memory management (garbage collection) used in Java does not play a major role in the divergent behaviour, as the algorithms were designed to minimise the required allocations, thus reducing the time spent on memory deallocations. However, on platform P3, the JVM represents a considerable bottleneck in the processing, and therefore, Java implementations of MwI are not suitable for weak platforms.

5. Conclusion

This chapter considers three different implementations of the Move with interleaving (MwI) transform with the aim to optimise the spent CPU time. Although MwI operates with the expected $O(n)$ time complexity, its CPU time usage differs considerably between various programming solutions. Although optimal implementations are necessary in embedded systems or computationally weaker platforms like Raspberry Pi, processing larger datasets can also be noticeably faster on modern personal computers. The first implementation utilised a standard vector to store the MwI status, the second incorporated a lookup table, while the third was based on a dynamic one-way connected list and a mechanism for reusing allocated memory chunks. All the implementations were encoded in C++ and Java.

Additionally, three different platforms were used for the experiments: MS Windows 11, Linux with the Manjaro distribution, and Raspberry Pi with Linux. The first two platforms were run on the same computer to ensure a fair comparison. The experiments involved 32 continuous-tone greyscale raster images with various resolutions and contexts. Their pixels were transformed into a sequence using a scan-line approach and then processed with MwI. The results show that careful implementation reduces the CPU time spent significantly. For example, the implementation based on the dynamic linked list was 1800% faster than the implementation using the standard vector. The lookup table accelerated the implementation, using a standard vector significantly; however, it did not reach the speed of the implementation with the dynamic linked list. Linux turned out to be slightly more efficient that Windows. As expected, the C++ implementations outperformed the Java implementations.

The Raspberry Pi platform cannot be compared directly to the two platforms running on a modern personal computer. However, the importance of careful implementation and the choice of programming language becomes even more evident. For example, the fastest implementation using a dynamic one-way connected linked list took 19 second if coded in C++, but one and a half minutes if coded in Java. The slowest implementation, using a standard vector, required 7 minutes and 32 seconds, while the same implementation coded in Java needed 1 hour and 42 minutes.

The features of MwI need further study. Firstly, the possibility of self-adaptation of the threshold t should be explored, based on some previous samples. Secondly, MwI and other string transformation techniques should be tested on larger alphabets, defined by, for example, high-quality raster images and audio data. In this case, list L

will contain 2^{16}, 2^{24}, or even 2^{32} elements, and its fast updating would represent a considerable challenge, especially if real-time MwI transform would be required.

Acknowledgements

This research was funded by the Slovenian Research and Innovation Agency under Research Project J2-4458, Research Programme P2-0041, and the Czech Science Foundation under Research Project 23-04622 L.

Author details

Borut Žalik[1*†], Aljaž Jeromel[1†], Ivana Kolingerová[2†], Niko Lukač[1†] and Blaž Repnik[1†]

1 Faculty of Electrical Engineering and Computer Science, University of Maribor, Maribor, Slovenia

2 Department of Computer Science and Engineering, University of West Bohemia, Plzeň, Czech Republic

*Address all correspondence to: borut.zalik@um.si

† These authors contributed equally.

References

[1] Žalik B, Strnad D, Podgorelec D, Kolingerová I, Lukač L, Lukač N, et al. A new transformation technique for reducing information entropy: A case study on greyscale raster images. Entropy. 2023;**25**(12):1591. DOI: 10.3390/e25121591

[2] Shannon CE. A mathematical theory of communication. AT&T Technical Journal. 1948;**27**(3):379-423. DOI: 10.1002/2Fj.1538-7305.1948.tb01338.x

[3] Cover TM, Thomas JA. Elements of Information Theory. 2nd ed. Hoboken: Wiley; 2006. DOI: 10.1002/047174882X

[4] Burrows M, Wheeler DJ. A Block-Sorting Lossless Data Compression Algorithm. Technical Report No. 124, Digital Systems Research Center; 1994. Available from: http://www.hpl.hp.com/techreports/Compaq-DEC/SRC-RR-124.pdf [Accessed: October 1, 2024]

[5] Adjeroh D, Bell T, Mukherjee A. The Burrows-Wheeler Transform: Data Compression, Suffix Arrays, and Pattern Matching. 2nd ed. New York: Springer Science + Business Media; 2008. DOI: 10.1007/978-0-387-78909-5

[6] Abel J. Post BWT stages of the Burrows-Wheeler compression algorithm. Software: Practice and Experience. 2010;**40**(9):751-777. DOI: 10.1002/spe.982

[7] Ryabko BY. Data compression by means of a 'book stack'. Problems of Information Transmission. 1980;**16**(4):265-269. Available from: https://zbmath.org/?q=an:0466.94007 [Accessed: October 1, 2024]

[8] Salomon D, Motta G. Handbook of Data Compression. 5th ed. London: Springer; 2010. DOI: 10.1007/978-1-84882-903-9

[9] Arnavut Z, Magliveras SS. Block sorting and compression. In: Storer JA, Cohn M, editors. Proceedings of the IEEE Data Compression Conference, DCC'97; 25–27 March 1997. Snowbird, Utah, USA: IEEE Computer Society Press; 1997. pp. 181-190. DOI: 10.1109/DCC.1997.582009

[10] Sayood K. Introduction to Data Compression. 4th ed. Waltham: Morgan Kaufman, Elsevier; 2012

[11] Clark BG. An efficient implementation of the algorithm "CLEAN". Atron Astrophy. 1980;**89**(3):377-378. Available from: https://ui.adsabs.harvard.edu/abs/1980A%26A....89..377C/abstract [Accessed: October 1, 2024]

[12] Hadjiconstantinou E, Christofides N. An efficient implementation of an algorithm for finding K shortest simple paths. International Journal of Network. 1999;**34**(2):88-101. DOI: 10.1016/j.tre.2019.11.013

[13] Dorado R, Gómez-Moreno A, Torres-Jiménez E, López-Alba E. An AHP application to select software for engineering education. Computer Applications in Engineering Education. 2014;**22**(2):200-208. DOI: 10.1002/cae.20546

[14] Goldberg AV. An efficient implementation of a scaling minimum-cost flow algorithm. Journal of Algorithms. 1997;**22**(1):1-29. DOI: 10.1006/jagm.1995.0805

[15] Shi G, Liu S, Chen P. A fast NURBS interpolation method for 3D ship hull

surface. Journal of Applied Sciences. 2013;**13**(12):2139-2145. DOI: 10.3923/jas.2013.2139.2145

[16] Lippert RA, Mobarry CM, Walenz BP. A space-efficient construction of the Burrows–Wheeler transform for genomic data. Journal of Computational Biology. 2005;**12**(7): 943-951. DOI: 10.1089/cmb.2005.12.943

[17] Žalik B. An efficient sweep-line Delaunay triangulation algorithm. Computer-Aided Design. 2005;**37**(10): 1027-1038. DOI: 10.1016/j.cad.2004.10.004

[18] Lukač N, Mongus D, Žalik B, Štumberger G, Bizjak M. Novel GPU-accelerated high-resolution solar potential estimation in urban areas by using a modified diffuse irradiance model. Applied Energy. 2024;**353**:122129. DOI: 10.1016/j.apenergy.2023.122129

[19] Devkota S, Aschwanden P, Kunen A, Legendre M, Isaacs KE. CcNav: Understanding compiler optimizations in binary code. IEEE Transactions on Visualization and Computer Graphics. 2021;**27**(2):667-677. DOI: 10.1109/TVCG.2020.3030357

[20] Chattopadhyay S. Compiler design. 2nd ed. New Delhi: PHI Learning; 2022

[21] Basford PJ, Johnston SJ, Perkins CS, Garnock-Jones T, Tso FP, Pezaros D, et al. Performance analysis of single board computer clusters. Future Generation Computer Systems. 2020; **102**:278-291. DOI: 10.1016/j.future.2019.07.040

[22] Khan N, Yaqoob I, Hashem IAT, Inayat Z, Ali WKM, Alam M, et al. Big data: Survey, technologies, opportunities, and challenges. Scientific World Journal. 2014;**2014**:712826. DOI: 10.1155/2014/712826

[23] Stunkel CB, Graham RL, Shainer G, Kagan M, Sharkawi SS, Rosenburg B, et al. The high-speed networks of the summit and sierra supercomputers. IBM Journal of Research and Development. 2020;**64**(3/4):3:1-3:10. DOI: 10.1147/JRD.2020.2967330

[24] Iserte S, Prades J, Reaño C, Silla F. Improving the management efficiency of GPU workloads in data centres through GPU virtualization. Concurrency and Computation: Practice and Experience. 2019;**22**(2):e5275. DOI: 10.1002/cpe.5275

[25] Khriji S, Benbelgacem Y, Chéour R, El Houssaini D, Kanoun O. Design and implementation of a cloud-based event-driven architecture for real-time data processing in wireless sensor networks. The Journal of Supercomputing. 2022;**78**: 3374-3401. DOI: 10.1007/s11227-021-03955-6

[26] Žlaus D, Mongus D. Efficient method for parallel computation of geodesic transformation on CPU. IEEE Transactions on Parallel and Distributed. 2020;**31**(4):935-947. DOI: 10.1109/TPDS.2019.2953057

[27] Qiu M, Ming Z, Li J, Gai K, Zong Z. Phase-change memory optimization for green cloud with genetic algorithm. IEEE Transactions on Computers. 2015; **64**(12):3528-3540. DOI: 10.1109/TC.2015.2409857

[28] Albers S. Energy-efficient algorithms. Communications of the ACM. 2010;**53**(6):86-96. DOI: 10.1145/1735223.1735245

[29] Rivest R. On self-organizing sequential search heuristics. Communications of the ACM. 1976; **19**(2):63-67. DOI: 10.1109/SWAT.1974.18

[30] Bentley JL, Sleator DD, Tarjan RE, Wei VK. A locally adaptive data

compression scheme. Communications of the ACM. 1986;**29**(4):320-330. DOI: 10.1145/5684.5688

[31] Dorrigiv R, López-Ortiz A, Munro JI. An application of self-organizing data structures to compression. In: Vahrenhold J, editor. Lecture Notes in Computer Science 5526Proceedings of the 8th International Symposium on Experimental Algorithms, SEA 2009, 3–6 June 2009, Dortmund, Germany. Berlin, Germany: Springer; 2009. pp. 137-148. DOI: 10.1007/978-3-642-02011-7_14

[32] Jelenković PJ, Radovanović A. Least-recently-used caching with dependent request. Theoretical Computer Science. 2004;**326**:293-327. DOI: 10.1016/j.tcs.2004.07.029

[33] Žalik B, Lukač N. Chain code lossless compression using move-to-front transform and adaptive run-length encoding. Signal Processing: Image Communication. 2014;**29**(1):96-106. DOI: 10.1016/j.image.2013.09.002

[34] Arnavut Z. Move-to-front and inversion coding. In: Storer JA, Cohn M, editors. Proceedings of the IEEE Data Compression Conference, DCC'97; 25–27 March 1997. Snowbird, Utah, USA: IEEE Computer Society Press; 1997. pp. 193-202. DOI: 10.1109/DCC.2000.838159

[35] Žalik B, Žalik M, Lipuš B. On move-to-front implementation. In: Proceedings on 27th International Conference on Circuits, Systems, Communications and Computers (CSCC), 19–22 July, 2023. Rhodes Island, Greece: IEEE Computer Society Press; 2023. pp. 43-47. DOI: 10.1109/CSCC58962.2023.00015

[36] Manber U, Myers G. Suffix arrays: A new method for on-line string search. SIAM Journal on Computing. 1993;**22**(5): 935-948. DOI: 10.1137/0222058

[37] Nong G, Zhang S, Chan WH. Two efficient algorithms for linear time suffix array construction. IEEE Transactions on Computers. 2011;**60**(10):1471-1484. DOI: 10.1109/TC.2010.188

[38] Kärkkäinen J, Sanders P, Burkhardt S. Linear work suffix array construction. Journal of the ACM. 2017; **53**(6):918-936. DOI: 10.1145/1217856.1217858

[39] Li Z, Li J, Huo H. Optimal In-place suffix sorting. Information and Computation. 2022;**285**(B):104818. DOI: 10.1016/j.ic.2021.104818

[40] Stroustrup B. The C++ Programming Language. 4th ed. Upper Saddle River: Addison-Wesley; 2013

[41] Cormen TH, Leiserson CE, Rivest RL, Stein C. Introduction to Algorithms. 3th ed. Cambridge: Massachusetts Institute of Technology; 2009

[42] Drepper U. What Every Programmer Should know about Memory. Available from: https://people.freebsd.org/~lstewart/articles/cpumemory.pdf [Accessed: October 1, 2024]

[43] Repnik B, Žalik B. Six Implementations of MwI Transform. Available from: https://gemma.feri.um.si/projects/slovene-national-research-projects/j2-4458-data-compression-paradigm-based-on-omitting-self-evident-information/software/ [Accessed: October 1, 2024]

[44] Schildt H. Java: The Complete Reference. 12th ed. New York: McGraw Hill; 2021

Chapter 6

Swarm Intelligence-Inspired Nanobiosensing for Tumor Microenvironment Targeting

Shaolong Shi and Tiangui Hu

Abstract

Early tumor detection, which plays a vital role in cancer treatment, is research. The tumor microenvironment (TME) closely related to the development of tumor lesions has shown great potential for the smart nanosystems-based tumor detection or therapy approaches. However, most of the nanosystems-based approaches focus on the functional enhancement of nanomaterials, with few breakthroughs in their transportation modes. Recently, we have proposed a novel early tumor-targeting framework named in vivo computation to improve the targeting efficiency by innovating the transportation modes of nanorobots from a computational perspective. In the framework, TME with unique physiological and pathological characteristics provides assisted information for the real-time path planning of nanorobots, which play the role of computing agents in the in vivo computational process. The main research point in this phase is computational strategy development with consideration of in vivo constraints. In this chapter, we summarize the computational strategies developed by classifying the in vivo computation procedures as external manipulable ones and autonomous ones, which are based on two prevailing propelling manners of nanorobots.

Keywords: swarm intelligence, nanorobots, tumor targeting, computational strategies, nanobiosensing

1. Introduction

Cancer as a leading cause of human mortality has gained extensive attention from researchers from various disciplines in the past decades [1]. Early tumor detection based on nanoparticles is increasingly important in the treatment of various cancers [2–4]. Smart nanosystems composed of biocompatible nanomaterials are responsive to external conditions such as heat, light, magnetic or electric field, and ultrasounds, as well as in vivo temperature, altered pH, ionic strength, etc., which guarantees their prevalent applications in tumor diagnosis and therapy [5, 6]. Furthermore, the specificities of tumor microenvironment (TME) such as abnormal pH, blood vessels, hormones, and growth factors make it aligned to develop specific functionalized nanoparticles for tumor sensing and targeting [7, 8].

Though nanoparticles have shown great development in the past decades, the breakthroughs are mainly in their structural design and functional modification without any improvement of the transportation mode. According to the statistical result, the transport efficiency (i.e., the proportion of agents that detect the tumor to the total number of agents deployed) is as low as 0.7% by using the human systemic circulation method [9]. To improve the targeting efficiency, nanorobots with the abilities of sensing, actuating, signaling, and navigating have emerged and experienced rapid development during the last decade [10]. For example, inspired by the bacterial flagellum in terms of swimming behavior, Ghosh and Fischer have developed the helical cobalt-covered glass propeller that can reach a speed of about 40 um/s [11]. Cheng et al. reported a kind of synthetic self-propelled micromotors that can realize long-range hydrodynamic communication, which offers promising possibilities for the development of intelligent cooperative nanorobot systems [12]. Alternatively, this method cannot be applied to targeting early microscopic tumors, as determining the tumor's location is a prerequisite that is unavailable prior to the detection process, posing a "chicken-or-egg causality dilemma."

To overcome the challenges above that cannot be handled by traditional biomedical technologies and improve the efficiency of TME targeting, we have proposed a novel paradigm named "in vivo computation (IVC)" by transforming the TME targeting problem into a computational problem [13]. Under the umbrella of IVC, some computational strategies classified as external manipulable ones and autonomous ones are developed with the consideration of in vivo constraints.

In this chapter, we will systematically introduce the structure of IVC and summarize the currently developed computational strategies for TME targeting.

2. In vivo computation framework

2.1 Computation modeling

As shown in **Figure 1**, we proposed the IVC framework to solve the problem of early tumor targeting that cannot be realized effectively by using the traditional methods [13]. As the emergence of tumor lesion will induce the specific environment (i.e., TME) around it, we proposed a concept of biological gradient field (BGF) to represent the characteristic of TME that can be used as aided knowledge for the design of computational strategies. Nanorobots classified as externally manipulable ones and autonomous ones are regulated by the computational strategies to finish tumor targeting with the effect of various in vivo constraints. In the framework, the tumor site is seen as the optimal solution of the computational problem; the vascular network of high-risk tissue is the search space; the nanorobots are employed as computing agents; and the TME-derivated BGF is viewed as the objective function that provides fitness evaluation for the agents.

The external imaging device is used to deduce the BGF landscape by monitoring the changes in the state of nanorobots. Computational strategies are developed to optimize the BGF landscape by employing the in vivo computing agents. The procedure of tumor targeting is interpreted as nanorobots updating their locations step by step under the guidance of computational strategies to find the tumor site. Thus, this in vivo optimization problem can be expressed as follows (for consistency and without loss of generality, the minimization operation is performed) [14]:

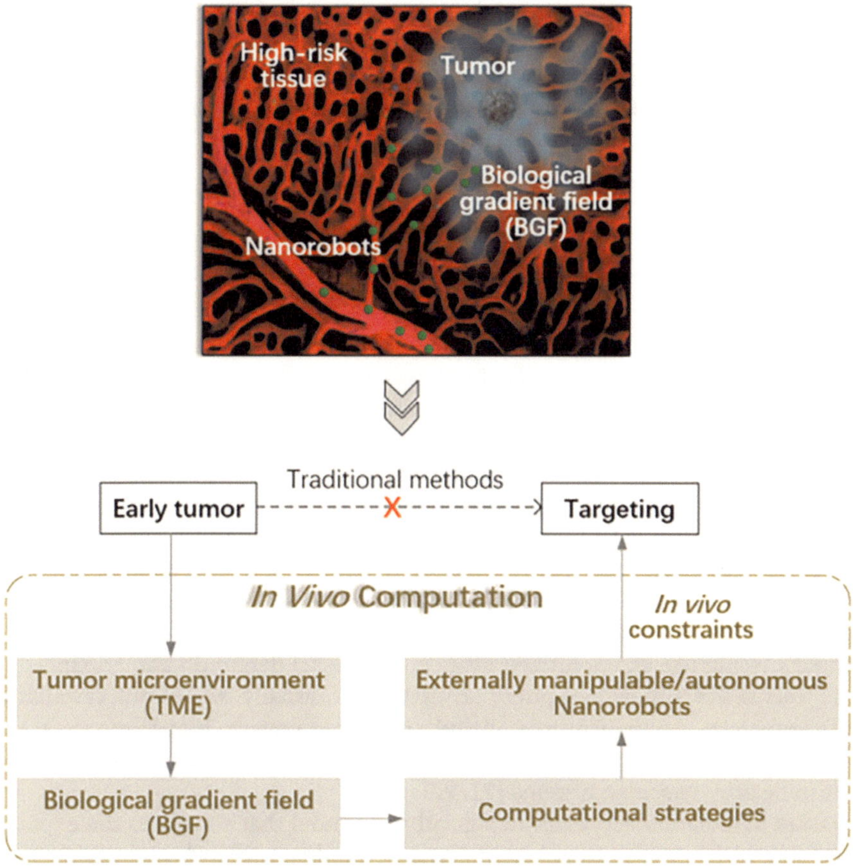

Figure 1.
Pictorial illustration of in vivo computation for tumor targeting and its system framework representation.

$$\begin{aligned} o(\boldsymbol{x};A) &= o_{app}(\boldsymbol{x};A) + o_{com}(\boldsymbol{x};A) \\ &= o_{int}(\boldsymbol{x}) + o_{dis}(\boldsymbol{x};A) + o_{com}(\boldsymbol{x};A), \end{aligned} \tag{1}$$

and

$$\boldsymbol{x}^* = arg\min_{\boldsymbol{x}} o(\boldsymbol{x};A), \boldsymbol{x} \in \mathbb{V}, \tag{2}$$

where $o_{app}(\boldsymbol{x};A)$ is the apparent objective function measured by agent A at location $\boldsymbol{x}$; $\boldsymbol{x} \in \mathbb{V}$ is a vector that represents the actual location of agent A in the vascular network $\mathbb{V}$; $o_{int}(\boldsymbol{x})$ is the intrinsic objective function at $\boldsymbol{x}$ independent of the presence or absence of A; $o_{dis}(\boldsymbol{x};A)$ is the extrinsic disturbance caused to the landscape due to the interaction between A and the search space $\mathbb{V}$; $o_{com}(\boldsymbol{x};A)$ is the correction factor accounting for the fitness of A due to the on-the-fly estimation of BGF; $\boldsymbol{x}$ is the global optimum that remains unchanged during the optimization process.

2.2 Representation of BGF

Given that current research has yet to establish a widely accepted and accurate quantitative model of BGF that can represent the diverse biological information surrounding tumors, we utilize several representative optimizable functions presented in **Table 1** in this study to capture the essential characteristics of BGF landscapes. All the BGF landscapes align generally with the qualitative observations reported in previous work and offer valuable insights into assessing the proposed method for exploring tumor boundaries [15]. Although additional landscapes could be utilized, it is practical to employ these as representations to test the proposed approach in this chapter.

2.3 Representation of BGF

After being injected into high-risk tissue, the nanorobots are controlled by an external actuating device to identify the tumor area within the human microvascular network [16]. Hence, the search space $\mathbb{V}$ encompasses the vasculature within the high-risk tissue, where capillaries function as suppliers of oxygen and nutrients necessary for the growth and proliferation of tumor cells [17]. On the other hand, the expansion of a tumor can prompt alterations in capillary morphology, and the attributes of the tumor's vascular network can mirror the evolving condition of the pathological tissue [18]. It has been shown that the processes of angiogenesis (generation of new blood vessels), vessel cooption (incorporation of existing vessels into the tumor's vascular system), and vessel regression transform the normal vascular network into a tumor-associated vasculature characterized by a central vessel depletion and enhanced angiogenesis at the tumor's periphery [19, 20]. Additionally, the tumor vasculature exhibits numerous unique features, including twisted vessels, loops, bypasses, highly variable distances between vessels, and extensive avascular zones, which are not present in healthy vascular systems [21, 22].

Invasion percolation serves as an algorithmic model that simulates the expansion of a network within a medium characterized by randomly distributed variations in strength, thereby accurately depicting the growth process of a tumor's vascular network [18, 23]. **Figure 2** illustrates the simulation of a vascular network in high-risk tissue using the invasion percolation algorithm. The tumor region, synthesized by the algorithm, is depicted as a white dotted circle with a diameter of 1 mm. The tumor vasculature undergoes a 40% regression. The area where angiogenesis takes place, at the tumor's periphery, is denoted by a square with a side length of 1.4 mm. The intercapillary distances are set to 0.1 mm for healthy tissue and 0.05 mm for the tumor periphery, respectively. The search space $\mathbb{V}$ is defined as a 10×10 mm area.

BGF landscapes	Search space
$(x^2+y^2+10)/85$	$x,y \in [-5,5]$
$1-\exp(-0.45(\lvert x\rvert+\lvert y\rvert))$	$x,y \in [-5,5]$
$-4/3\exp(-\sqrt{2x^2+2y^2}/10)+e/15+4/3-1/15\exp(1/2(\cos(2\pi x)+\cos(2\pi y)))$	$x,y \in [-5,5]$
$1/100[-10\cos(2\pi x)-10\cos(2\pi y)+x^2+y^2]+0.35$	$x,y \in [-5,5]$
$0.99-0.99\cos(3x)\cos(3y)\times\exp[-(9x^2+9y^2)]$	$x,y \in [-5,5]$
$13/1500(x^2+y^2)+6/375xy+1/6$	$x,y \in [-5,5]$
$(2x^2+y^2+xy)/75-7/3125x^4+8/140625x^6+1/6$	$x,y \in [-5,5]$

Table 1.
Representations of BGF landscapes.

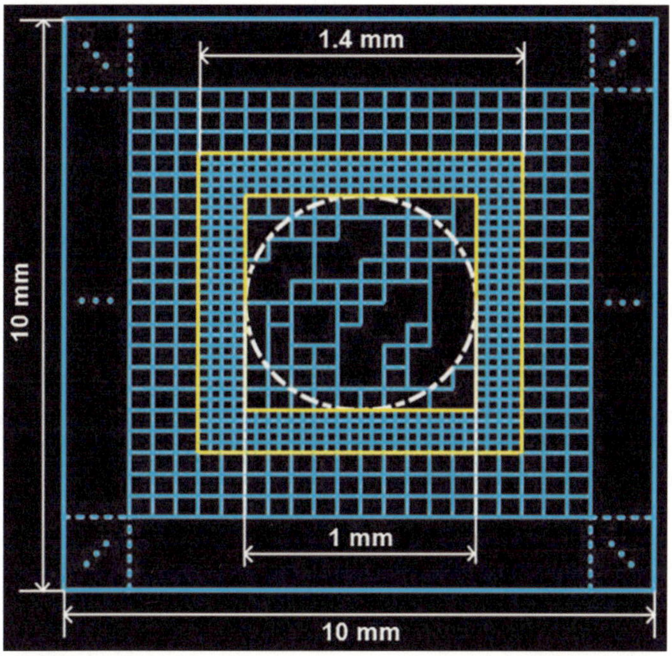

Figure 2.
In vivo computation involving the tumor vascular network simulated by the invasion percolation algorithm utilizes a search space $\mathbb{V}$ with dimensions of 10 mm × 10 mm. Within this space, the tumor periphery is delineated by a square of 1.4 mm × 1.4 mm, and the tumor itself is represented by a white dotted circle measuring 1 mm in diameter.

3. In vivo computational strategies

3.1 Weak priority evolution strategy (WP-ES)

In classic evolutionary computation, entities continually evolve based on fitness scores derived from a cost function within the search space, allowing diverse entities to progress in various directions during a single iteration. Unlike this process, the updating of nanorobot positions in tumor targeting is limited to a single, common direction due to the constraint imposed by the external, uniform magnetic field.

In our society, greater legal and ethical protections are often granted to vulnerable groups, promoting harmony, fairness, and progress within the community. From an optimization perspective, improving the performance of the lowest-performing agent can substantially positively impact the overall group fitness. Motivated by this observation, we introduce the Weak Priority Evolution Strategy (WP-ES) in this section. Herein, the weak priority implies that the agent exhibiting the poorest performance is always given precedence for activation and tracking in each iteration.

The WP-ES contains the following steps:

1. The agents operate in the imaging and tracking mode, and their fitness in the search space is evaluated according to the representative BGF landscapes shown in Section 2.2.

2. In the next moment, the updating direction of each agent is calculated according to a certain optimization algorithm, and the lowest-ranked agent (i.e., w_1) that has the worst fitness of the swarm is selected as shown in **Figure 3(a)**.

3. The actuating direction of the external magnetic field is adjusted according to the updating direction of agent w_1. Thus, agent w_1 works in the intended actuating mode, and all the other agents work in the unintended actuating mode with a uniform direction, as shown in **Figure 3(b)**.

4. All the agents work in the imaging and tracking mode, and their locations are estimated, and the same process as step (1) is repeated. Then the worst agent (i.e., w_2) in the next iteration is selected as shown in **Figure 3(c)**. The iteration process continues until some stopping criteria are met.

3.2 Exponential evolution mechanism

In optimization society, a search process that initially emphasizes enhanced exploration and subsequently focuses on heightened exploitation typically yields excellent results. Given the resemblance between in vivo computation and traditional optimization in their computational processes, it is crucial to initiate the search with a large step size to expedite the process and then switch to a smaller step size toward the end to boost the likelihood of precise TME targeting.

Therefore, we utilize an exponential evolution mechanism for in vivo computation by progressively reducing the step size of nanorobots in an exponential manner during the actuation phase.

$$d(t_k) = d_0 \cdot e^{-\alpha \cdot t_k / T} \tag{3}$$

As illustrated in Eq. (3), $d(t_k)$ represents the movement increment of the nanorobot when it operates in the kth actuation mode; the initial step size is set as d_0; the descent factor is denoted by α, and the maximum allowed operation time is set to T. As the operating time progresses, $d(t_k)$ will decrease according to this evolution mechanism. Moreover, since nanorobots can only traverse the search space in one direction, unlike the bidirectional search conducted by computing agents in

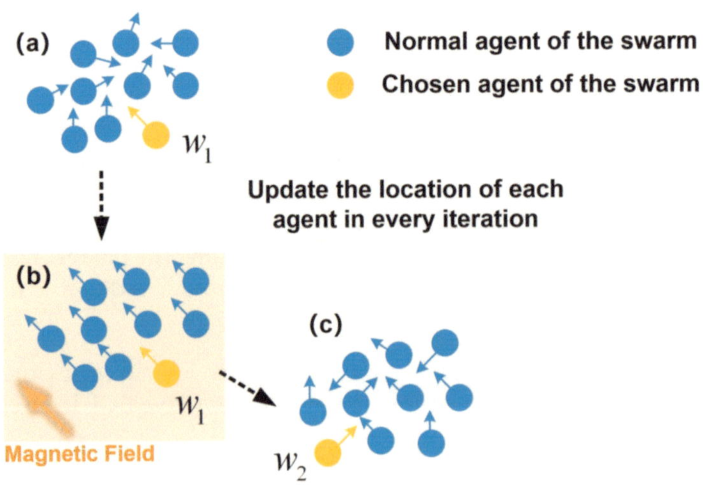

Figure 3.
The iterative process of agent evolution in a single cycle.

mathematical optimization, the ideal configurations for d_0 and α are expected to vary between in vivo computation and traditional swarm intelligence optimization.

It is important to emphasize that the aim of this research is to devise an effective tumor targeting strategy. Initially, we establish analogies between the tumor targeting process and optimization problems, enabling us to leverage a variety of optimization algorithms from the evolutionary computation domain, such as Gravitational Search Algorithm (GSA) and Particle Swarm Optimization (PSO), to inspire innovative tumor targeting approaches. We refer to this as "in vivo computation," where nanorobots serve as "computing agents" to conduct a "natural" search for solutions (i.e., TME targeting). Our objective is to enhance TME targeting outcomes by applying the proposed strategy while accounting for the physical constraints specific to in vivo computation, rather than developing a novel algorithm for traditional optimization problems.

3.3 Tumor boundary exploration strategy

Resulting from the response of nanorobots with the BGF landscape, the nanorobots will coordinate migration, which is the fundamental motion mechanism of autonomous nanorobots. Assuming N nanorobots are deployed in the search space for tumor targeting with their locations represented by $x_1, x_2, \cdots, x_N$, the values of sensed BGF with them are $f(x_1), f(x_2), \cdots, f(x_N)$. Furthermore, K neighbors of a nanorobot are selected according to the nearest neighbor criterion. The specific nanorobot interacts with its neighboring nanorobots in such a way that it is attracted toward a neighbor with a high BGF value and repelled from one with a low BGF value, facilitating coordinated movement. Hence, the direction of migration for the particular nanorobot can be determined using the following calculation:

$$\boldsymbol{u} = \frac{\boldsymbol{x}_n - \boldsymbol{x}_k}{\|\boldsymbol{x}_n - \boldsymbol{x}_k\|} \tag{4}$$

where $\boldsymbol{x}_n$ is the location of a specific nanorobot in the swarm with $i = 1, 2, \cdots, N$, and $\boldsymbol{x}_k$ is the location of its neighbor with $i = 1, 2, \cdots, K$.

The effects of the neighbors on the specific nanorobot are:

$$\mathcal{H} = \sum_{i=1}^{K} \frac{\Delta f_{n,k}}{\left[\|\boldsymbol{x}_n - \boldsymbol{x}_k\| / d_0\right]^l} \boldsymbol{u}, \tag{5}$$

where $\Delta f_{n,k} = f(\boldsymbol{x}_k) - f(\boldsymbol{x}_n)$ is the difference of BGF values perceived by the specific nanorobot and its neighbor; d_0 is a normalization factor; l is the path loss index.

The location update of the nanorobot with cooperation mechanism is:

$$\boldsymbol{x}_{n,t+1} = \boldsymbol{x}_{n,t} + \gamma \bullet \mathcal{H}, \tag{6}$$

where γ is the unit step size. Subsequently, each nanorobot within the swarm adjusts its position based on this cooperative mechanism.

As nanorobots move autonomously, they generate powerful and consistent phoretic flows, emitting robust hydrodynamic signals. These signals trigger an "approach-hit-and-run" response among neighboring nanorobots within the perception distance, effectively creating a competitive mechanism where nanorobots exhibit strong repulsion toward each other within a certain range.

The effect of repulsion can be calculated as follows:

$$T(t) = \sum_{i=1}^{K} \left[\frac{q(t)}{8\pi\lambda \|\boldsymbol{x}_n - \boldsymbol{x}_k\|^3} \bullet \Gamma(\|\boldsymbol{x}_n - \boldsymbol{x}_k\| - R) + \frac{q(t)}{8\pi\lambda R^3} \bullet \Gamma(R - \|\boldsymbol{x}_n - \boldsymbol{x}_k\|) \right] \boldsymbol{u} \quad (7)$$

where $\Gamma(x) = \begin{cases} x, & x \geq 0 \\ 0, & x < 0 \end{cases}$; $q(t)$ is the polarization intensity of the nanorobot; λ is the viscosity coefficient of the blood; R is the radius of repulsion.

The location update of the nanorobot with competition mechanism is:

$$\boldsymbol{x}_{n,t+1} = \boldsymbol{x}_{n,t} + \gamma \bullet T(t), \quad (8)$$

Thus, the nanorobots move in the search space to target the tumor and explore the tumor region by using the cooperation and competition mechanisms autonomously, which consist of the tumor boundary exploration strategy.

4. Numerical experiments

We use several numerical examples to elaborate on the TME targeting performance of the proposed evolution strategies and compare them to that of the traditional method (i.e., brute-force search). In the experiments, 12 agents are employed in the search space; the initial deployment region is -5 mm $\leq x, y \leq -4$ mm and the speed of agents is 150μm/s. The maximum iteration is 60, and the total number of simulation runs is 1000. The cancer is denoted by a circle of radius 0.5 mm in the center of the search space. Each agent reaching the cancer will stop moving, and the other agents will continue searching until the iteration reaches the maximum number. The BGF landscapes presented in Section 2.2 are applied to synthesize different levels of TME targeting difficulty. Two performance metrics are considered, namely, the probability of cancer detection Pd (i.e., the ratio of cancer detected successfully in all the simulation runs) and the percentage of agents delivered to the cancer focus η (i.e., the ratio of agents detecting the cancer in all the employed agents).

Figure 4 shows the typical curves of the employed agents' average fitness over iterations in the representative BGF landscape by using five different search algorithms: brute-force (B-F) search and S-ES, WP-ES, SP-ES, and R-ES-based swarm intelligence. Please note that S-ES, SP-ES, and R-ES are all classical evolution strategies used as benchmarks. As can be seen from the figure, the average fitness of agents relying on WP-ES reaches the maximum at the 22th iteration. Its value is higher than any other curve in any iteration, meaning that WP-ES performs better than other search algorithms in this scenario.

To compare the search algorithms quantitatively, we have carried out the simulation for 1000 runs and presented the histograms of the results in **Figure 5**. It can be seen that WP-ES yields a detection ratio of Pd = 100%, which is the same as that of B-F (Pd = 100%, **Figure 5(a)**) and is similar to that of S-ES (Pd = 99.2%, **Figure 5(b)**) and higher than that of SP-ES (Pd = 0, **Figure 5(d)**) and R-ES (Pd = 76.1%, **Figure 5 (e)**). In terms of the targeting efficiency, WP-ES achieves a much higher value of η = 84.8% compared to that of B-F (η = 56.1%), S-ES (η = 59.7%), SP-ES (η = 0), and R-ES (η = 36.3%).

To demonstrate the effectiveness of the exponential evolution mechanism, we carry out the experiment with d_0= 1, α= 3. **Figure 6(a)** shows the typical curve of the

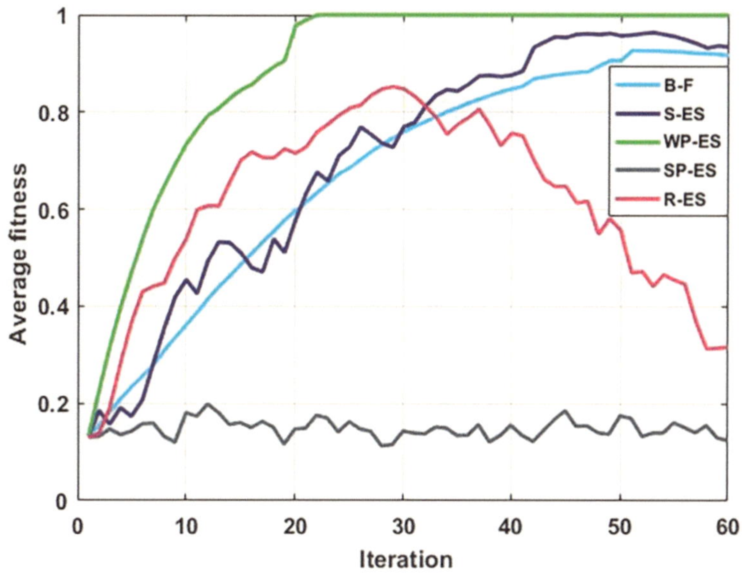

Figure 4.
Average fitness of the agents input over iterations in the BGF landscape.

agents' average fitness over the search time for BGF landscape by using the exponential evolution mechanism in Section 3.2. It shows that the average fitness of agents reaches the maximum, around 0.9, at 200 s. **Figure 6(b)** shows that six agents can detect the tumor with the trajectory of the swarm center passing through the tumor region, and one agent is dysfunctional at 500 s. **Figure 7(a)** and **(b)** show the histograms of the minimum targeting time and the agents targeting the tumor by using the exponential evolution mechanism and the tumor targeting results, where τ = 172.9 s, η= 37.1%, Pd = 96.4%, and ζ = 16%. Thus, it is striking that the exponential evolution mechanism-based in vivo computational algorithm performs better for TME targeting.

To characterize the result of tumor boundary exploration by using autonomous computational nanobiosensing technology, we carry out several numerical simulation experiments. In the experiments, 50 agents are utilized as autonomous nanorobots in the search space, where a circle region with a radius of 5 mm in the center represents the tumor region. The number of neighbors K is 10, and the movement step size γ is 0.05 mm.

Figure 8 shows the procedure of tumor boundary in a representative BGF landscape (other landscapes have similar results, which will not be overexpressed here). As shown in **Figure 8(a)**, the initial distribution of nanorobots is in the bottom left corner of the search space. All the nanorobots will move with the effect of cooperation mechanisms and competition mechanisms. With the process going on, the nanorobots will detect the tumor region more and more, as shown in **Figure 8(b)**. **Figure 8(c)** shows the final result of the tumor boundary exploration with a state of balance reached by the cooperation mechanism and competition mechanism. All the nanorobots are located dispersively in the tumor region, and some are distributed on the tumor boundary, indicating a desired exploration result of the tumor boundary.

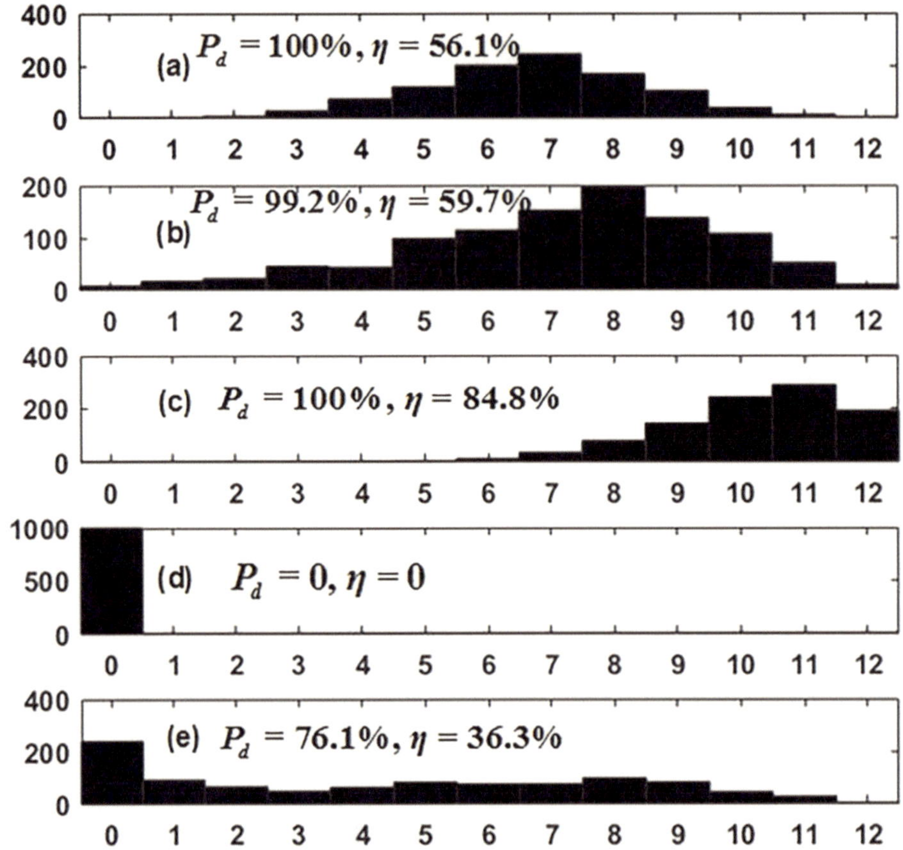

Figure 5.
Histogram of the quantity of agents located at the cancer using different search algorithms and their performance: (a) brute-force search, (b) S-ES-based swarm intelligence, (c) WP-ES-based swarm intelligence, (d) SP-ES-based swarm intelligence, and (e) R-ES-based swarm intelligence.

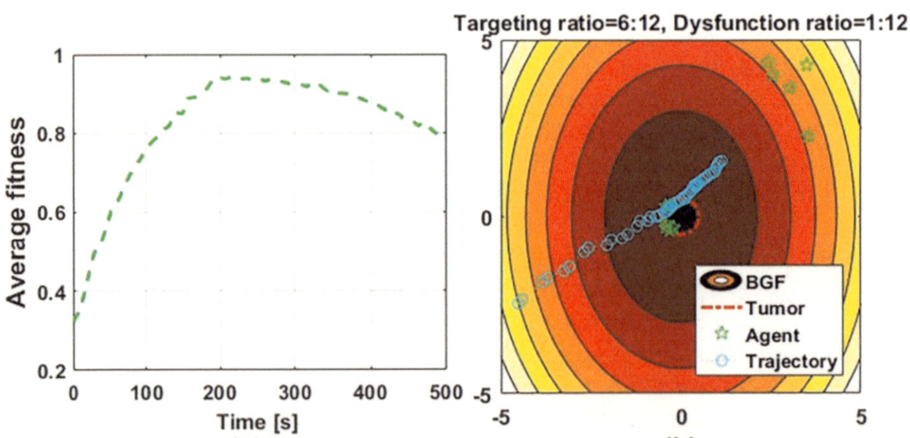

Figure 6.
TME targeting results for BGF landscape by using exponential evolution mechanism. (a) Average fitness of the agents input during the search time; (b) final locations of the agents input in the search space after one run; and the trajectory of the agent swarm center.

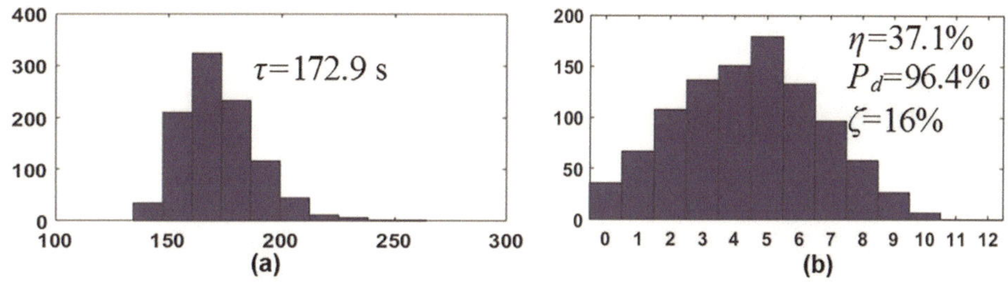

Figure 7.
Histograms of the TME targeting results by using the exponential evolution mechanism. (a) The minimum targeting time; (b) quantity of agents targeting the tumor.

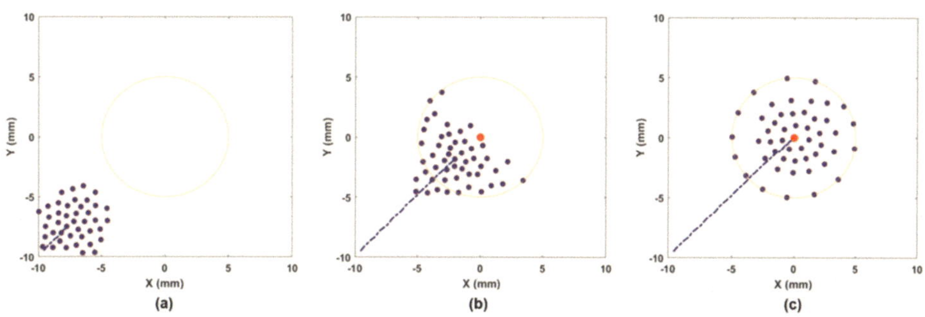

Figure 8.
Distributions of nanorobots at three different moments: (a) the initial moment, (b) the intermediate moment, and (c) the terminal moment.

5. Conclusion

In this chapter, we summarized the swarm intelligence-based computational strategies developed for the control and tracking of nanorobots during the TME targeting process.

The proposed WP-ES is used to overcome the physical constraint of an external uniform magnetic field for the effective actuating of nanorobots. The exponential evolution strategy is used to improve the targeting efficiency of nanorobots. Furthermore, we proposed the tumor boundary exploration strategy, which contains the cooperation and competition mechanisms of nanorobots for the exploration of tumor lesion areas. The numerical experiments demonstrated the effectiveness of the proposed strategies.

Acknowledgements

This work is supported by the National Natural Science Foundation of China (Grant No. 62102071, 62472322) and the Municipal Government of Quzhou (Grant No. 2023D031, 2023D021).

Conflict of interest

The authors declare no conflict of interest.

Author details

Shaolong Shi* and Tiangui Hu
Yangtze Delta Region Institute (Quzhou), University of Electronic Science and Technology of China, Quzhou, China

*Address all correspondence to: shishaolong@csj.uestc.edu.cn

References

[1] G. B. of Disease 2019 Cancer Collaboration. Cancer incidence, mortality, years of life lost, years lived with disability, and disability adjusted life years for 29 cancer groups from 2010 to 2019: A systematic analysis for the global burden of disease study 2019. JAMA Oncology. 2022;**8**(3):420-444

[2] Crosby D, Bhatia S, Brindle KM, Coussens LM, Dive C, Emberton M, et al. Early detection of cancer. Science. 2022;**375**(6586):eaay9040

[3] Anselmo AC, Mitragotri S. Nanoparticles in the clinic: An update. Bioengineering & Translational Medicine. 2019;**4**(3):e10143

[4] Brigger I, Dubernet C, Couvreur P. Nanoparticles in cancer therapy and diagnosis. Advanced Drug Delivery Reviews. 2012;**64**:24-36

[5] Kwon EJ, Lo JH, Bhatia SN. Smart nanosystems: Bio-inspired technologies that interact with the host environment. Proceedings of the National Academy of Sciences. 2015;**112**(47):14460-14466

[6] Singh S, Sirbaiya A, Mishra A. Bioinspired smart nanosystemsin advanced therapeutic applications. Pharmaceutical Nanotechnology. 2019; **7**(3):246-256

[7] Arneth B. Tumor microenvironment. Medicina. 2019;**56**(1):15

[8] Anderson NM, Simon MC. The tumor microenvironment. Current Biology. 2020;**30**(16):R921-R925

[9] Wilhelm S, Tavares AJ, Dai Q, Ohta S, Audet J, Dvorak HF, et al. Analysis of nanoparticle delivery to tumours. Nature Reviews Materials. 2016;**1**(5):1-12

[10] Arvidsson R, Hansen SF. Environmental and health risks of nanorobots: An early review. Environmental Science: Nano. 2020; **7**(10):2875-2886

[11] Ghosh A, Fischer P. Controlled propulsion of artiffcial magnetic nanostructured propellers. Nano Letters. 2009;**9**(6):2243-2245

[12] Cheng Y, Mou F, Yang M, Liu S, Xu L, Luo M, et al. Long-range hydrodynamic communication among synthetic self-propelled micromotors. Cell Reports Physical Science. 2022;**3**(2): 100739

[13] Shi S, Shariff N, Cheang UK, Chen Y. Perspective: Computational nanobiosensing. IEEE Transactions on Nanobioscience. 2019;**19**(2):267-269

[14] Shi S, Chen Y, Ding J, Liu Q, Zhang Q. Dynamic In vivo computation for learning-based Nanobiosensing in time-varying biological landscapes. IEEE Transactions on Evolutionary Computation. 2022;**27**(4): 1100-1114

[15] Peng SJ, Xiao FF, Chen MW, et al. Tumor-microenvironment responsive nanomedicine for enhanced cancer immunotherapy. Advanced Science. 2022;**9**(1):2103836

[16] Chen Y, Ali M, Shi S, Cheang UK. Biosensing-by-learning direct targeting strategy for enhanced tumor sensitization. IEEE Transactions on Nanobioscience. 2019;**18**(3):498-509

[17] Lorthois S, Cassot F. Fractal analysis of vascular networks: Insights from morphogenesis. Journal of Theoretical Biology. 2010;**262**(4):614-633

[18] Gazit Y, Baish JW, Safabakhsh N, Leunig M, Baxter LT, Jain RK. Fractal characteristics of tumor vascular architecture during tumor growth and regression. Microcirculation. 1997;**4**(4): 395-402

[19] Holash J, Maisonpierre P, Compton D, Boland P, Alexander C, Zagzag D, et al. Vessel cooption, regression, and growth in tumors mediated by angiopoietins and vegf. Science. 1999;**284**(5422):1994-1998

[20] Carmeliet P, Jain RK. Angiogenesis in cancer and other diseases. Nature. 2000;**407**(6801):249

[21] Baish JW, Jain RK. Fractals and cancer. Cancer Research. 2000;**60**(14): 3683-3688

[22] Rieger H, Welter M. Integrative models of vascular remodeling during tumor growth. Wiley Interdisciplinary Reviews: Systems Biology and Medcine. 2015;**7**(3):113-129

[23] Baish JW, Gazit Y, Berk DA, Nozue M, Baxter LT, Jain RK. Role of tumor vascular architecture in nutrient and drug delivery: An invasion percolation-based network model. Microvascular Research. 1996;**51**(3): 327-346